INTRODUCTION TO CIVIL ENGINEERING

AUTHOR(s)

Prof. Manabendra Saharia
Assistant Professor,
Department of Civil Engineering,
Hauz Khas, IIT Delhi

Prof. Nagendra R. Velaga
Professor,
Department of Civil Engineering,
IIT Bombay

REVIEWER

Dr. Sumant Kulkarni
Associate Professor,
Civil Engineering (CE), D.Y. Patil University Pune Ambi,
Talegaon Dabhade, Pune, Maharashtra

KHANNA BOOK PUBLISHING CO. (P) Ltd.
4C/4344, Ansari Road, Darya Ganj,
New Delhi, India

All India Council for Technical Education
Nelson Mandela Marg, Vasant Kunj,
New Delhi, India

Dear Readers,

To prevent the piracy, this book is secured with HIGH SECURITY HOLOGRAM on the front title cover. In case you don't find the hologram on the front cover title, please write us to at contact@khannabooks.com or whatsapp us at +91-99109 09320 and avail special gift voucher for yourself.

Specimen of Hologram on front Cover title: 

Moreover, there is a SPECIAL DISCOUNT COUPON for you with EVERY HOLOGRAM.

How to avail this SPECIAL DISCOUNT:

Step 1: Scratch the hologram

Step 2: Under the scratch area, your "coupon code" is available

Step 3: Logon to www.khannabooks.com

Step 4: Use your "coupon code" in the shopping cart and get your copy at a special discount

Step 5: Enjoy your reading!

ISBN: 978-93-5538-704-2

Book Code: UG406EN

Introduction To Civil Engineering

by Manabendra Saharia, Nagendra R. Velaga

First Edition: 2026

Originally Published by:

All India Council for Technical Education, Ministry of Education, GOI

Exclusively Published & Distributed by:

Khanna Book Publishing Co. (P) Ltd.

4C/4344, Ansari Road, Daryaganj,
New Delhi (INDIA)

Visit us at: **www.khannabooks.com**

Write us at: **contact@khannabooks.com**

To view complete list of books, Please scan the QR Code: KPH

Printed in Bharat.

प्रो. टी. जी. सीताराम
अध्यक्ष
Prof. T. G. Sitharam
Chairman

सत्यमेव जयते

आज़ादी का अमृत महोत्सव

अखिल भारतीय तकनीकी शिक्षा परिषद्
(भारत सरकार का एक सांविधिक निकाय)
(शिक्षा मंत्रालय, भारत सरकार)
नेल्सन मंडेला मार्ग, वसंत कुंज, नई दिल्ली-110070
दूरभाष : 011-26131498
ई-मेल : chairman@aicte-india.org

ALL INDIA COUNCIL FOR TECHNICAL EDUCATION
(A STATUTORY BODY OF THE GOVT. OF INDIA)
(Ministry of Education, Govt. of India)
Nelson Mandela Marg, Vasant Kunj, New Delhi-110070
Phone : 011-26131498
E-mail : chairman@aicte-india.org

FOREWORD

Engineers are the backbone of any modern society. They are the ones responsible for the marvels as well as the improved quality of life across the world. Engineers have driven humanity towards greater heights in a more evolved and unprecedented manner.

The All India Council for Technical Education (AICTE), have spared no efforts towards the strengthening of the technical education in the country. AICTE is always committed towards promoting quality Technical Education to make India a modern developed nation emphasizing on the overall welfare of mankind.

An array of initiatives has been taken by AICTE in last decade which have been accelerated now by the National Education Policy (NEP) 2020. The implementation of NEP under the visionary leadership of Hon'ble Prime Minister of India envisages the provision for education in regional languages to all, thereby ensuring that every graduate becomes competent enough and is in a position to contribute towards the national growth and development through innovation & entrepreneurship.

One of the spheres where AICTE had been relentlessly working since past couple of years is providing high quality original technical contents at Under Graduate & Diploma level prepared and translated by eminent educators in various Indian languages to its aspirants. For students pursuing 2nd year of their Engineering education, AICTE has identified 87 books, which shall be translated into 12 Indian languages - Hindi, Tamil, Gujarati, Odia, Bengali, Kannada, Urdu, Punjabi, Telugu, Marathi, Assamese & Malayalam. In addition to the English medium, books in different Indian Languages are going to support the students to understand the concepts in their respective mother tongue.

On behalf of AICTE, I express sincere gratitude to all distinguished authors, reviewers and translators from the renowned institutions of high repute for their admirable contribution in a record span of time.

AICTE is confident that these outcomes based original contents shall help aspirants to master the subject with comprehension and greater ease.

(Prof. T. G. Sitharam)

FOREWORD ABOUT THE BOOK

I am delighted to introduce this book, "Introduction to Civil Engineering," to the next generation of aspiring civil engineers. Civil Engineering is the mother of engineering, and a student is expected to acquire a variety of skills such as excellent scientific and mathematical temper, sound health, common sense, excellent interpersonal and communication skills.

Domain of civil engineering can be divided in many sub-areas such as Structural engineering, Transportation engineering, Hydraulics and Water Resources engineering, Geotechnical engineering, Geomatics engineering, Environmental engineering, Port and Harbour engineering, etc.

Structural engineering is concerned with the analysis and design of structures such as dams, weirs, bridges, skyscrapers, tunnels, and so on. The structures should have desired strength, durability, and stability; they should be aesthetically appealing, and should be economical. Transportation engineering focuses on the planning, design, construction, and maintenance of transportation systems and infrastructure, including roads, highways, airports, and railways. Hydraulics and Water Resources engineering deals with the principles of water and sediment/pollutant flow, and development and management of water resources for meeting needs of municipal and industrial water supply, irrigation, hydropower generation, flood and drought control, and recreation. It also includes water supply, treatment, and distribution systems. Geotechnical engineering deals with the properties and behavior of soils and rocks and finds applications in the design and construction of foundations, slopes, and tunnels. Environmental engineering focuses on the protection of human health and environment and involves the application of scientific and engineering principles to prevent, control, and remediate environmental problems. Geomatics engineering deals with surveying and measurements of terrestrial features using earth- or space-based instruments/sensors. Port and Harbor engineering is concerned with planning, design, construction, and running of marine-based transport systems. Sustainability is an over-arching concept of civil engineering since the solutions provided should be environment-friendly, socially, and culturally acceptable, economically viable, and sustainable.

This book provides a solid foundation in the core disciplines of civil engineering. It also includes numerous QR codes that help readers to access additional resources and interactive contents, such as videos, animations, and simulations. The book also contextualizes civil engineering in India, highlighting the challenges and opportunities in this rapidly developing country.

I am confident that as a first course in a civil engineering undergraduate program, this book will be an invaluable resource for students who are eager to learn about the exciting and rewarding field of civil engineering. It will provide them with the knowledge and skills they need to tackle the complex and pressing challenges facing our world today, and to help build a more sustainable and livable future for all.

Sharad K Jain
Indian Institute of Technology Roorkee
Former Director, National Institute of Hydrology

ACKNOWLEDGEMENT

The authors are grateful to the authorities of AICTE, particularly Prof. T. G. Sitharam, Chairman; Prof. M. P. Poonia, Vice-Chairman; Prof. Rajive Kumar, Member-Secretary and Dr Amit Kumar Srivastava, Director, Faculty Development Cell for their planning to publish the book on "Introduction to Civil Engineering." We sincerely acknowledge the valuable contributions of the reviewer of the book, Dr. Sumant Kulkarni, REVA University.

This book is an outcome of various suggestions of AICTE members, experts and authors who shared their opinion and thought to further develop engineering education in our country. Acknowledgements are due to the different professionals in this field whose published books, review articles, papers, photographs, footnotes, references, and other valuable information enriched us at the time of writing the book.

Manabendra Saharia dedicates this book to his late mother, Ms. Nilakhi Sahariah. He would also like to acknowledge the support of his wife, Prof. Shrutidhara Sarma, and his father, brother, father-in-law, and mother-in-law. He acknowledges the extraordinary debt he owes to his well-wishers over the years: Prof. Rajib Bhattacharjya, Prof. Sharad K. Jain, Prof. Parthajit Roy, Prof. Parthasarathi Choudhury, Prof. Ramgopal Rao, Prof. G. V. Ramana, Prof. A.K. Nema, Prof. Sumedha Chakma, Prof. DR Kaushal, Prof. Rangan Banerjee, Prof. Pierre Kirstetter, Dr. Jonathan J. Gourley, and Prof. Yang Hong. Nagendra R. Velaga dedicates this book to his parents (Velaga Subbarao and Velaga Rama devi) and family members Dr. Sree Lalitha, Jaya, and Shiv. This work is also dedicated to his teachers, who have spent a significant amount of their life in building him.

Finally, we would like to acknowledge the heroic efforts of our students at IIT Delhi and IIT Bombay, without which writing this book would not have been possible.

Prof. Manabendra Saharia, IIT Delhi

Prof. Nagendra R. Velaga, IIT Bombay

PREFACE

The book titled "Introduction to Civil Engineering" is an outcome of the rich experience of our teaching of basic and advanced Civil Engineering subjects at IIT Delhi and IIT Bombay. The initiation of writing this book is to expose the extraordinary breadth of civil engineering to undergraduate students as a first course in civil engineering. Keeping in mind the purpose of wide coverage as well as to provide essential supplementary information, we have included the topics recommended by AICTE, in a very systematic and orderly manner throughout the book. Efforts have been made to explain the fundamental concepts of the subject in the simplest possible way.

During the process of preparation of the book, we have considered various standard textbooks and reports. While preparing the different sections, emphasis has also been laid to provide an overview of a vast array of civil engineering topics, but also delve a little deep into their applications in the real world. We have made extensive use of illustrations and examples to enrich this book for proper understanding of the related topics. Under the common title "Introduction to Civil Engineering," we cover 6 units dealing with the various subfields of Civil Engineering. They include structural engineering, water resources engineering, geotechnical engineering, transportation engineering, etc. We also provide a chapter on practical applications where we provide examples from different areas of civil engineering for classroom practice as well as stoking the personal interests of students. We have inserted QR codes into every chapter to point towards resources that students might want to explore beyond the textbook.

As far as the present book is concerned, it will allow civil engineering students to appreciate the breadth and depth of civil engineering as a topic and prepare them for upper-level courses.

We sincerely hope that the book will inspire the students to learn and discuss the ideas behind basic principles of civil engineering and will surely contribute to the development of a solid foundation of the subject. We would be thankful to all beneficial comments and suggestions which will contribute to the improvement of the future editions of the book. It gives us immense pleasure to place this book in the hands of the teachers and students. It was indeed a big pleasure to work on different aspects covered in the book.

Authors

Chapter-wise presentations for this book have been uploaded here for the convenience of students and educators

https://introtocivilengg.wixsite.com/book

CONTRIBUTORS TO THIS BOOK

Dr Nishant M. Pawar Postdoctoral Research Fellow Indian Institute of Technology Bombay	**Monik Gupta** PhD Candidate Indian Institute of Technology Bombay
Priyam Deka PhD Candidate Indian Institute of Technology Delhi	**Rohan Bargoti** M.Tech. Scholar Indian Institute of Technology Delhi
Ved Prakash PhD Candidate Indian Institute of Technology Delhi	**Dr. Satyadhrik Sharma** Department of Materials, Mechanics, Management & Design Technische Universiteit Delft
Shashank M.Tech. Scholar Indian Institute of Technology Delhi	**Roohmoney** PhD Candidate Indian Institute of Technology Delhi
Anagha P. PhD Candidatc Indian Institute of Technology Delhi	**Bhanu Magotra** PhD Candidate Indian Institute of Technology Delhi
Ravi Raj PhD Candidate Indian Institute of Technology Delhi	**Sai Kiran Kuntla** PhD Candidate Indian Institute of Technology Delhi
Nirdesh Kumar Sharma PhD Candidate Indian Institute of Technology Delhi	**Prateek Sharma** PhD Candidate Indian Institute of Technology Delhi
Suneet Bansal M.Tech. Scholar Indian Institute of Technology Delhi	**Prof. Prashanth Vangla** Assistant Professor Dept. of Civil Engineering, IIT Delhi
Prof. Sahil Bansal Assistant Professor Dept. of Civil Engineering, IIT Delhi	**Prof. Deepanshu Shirole** Assistant Professor Dept. of Civil Engineering, IIT Delhi

OUTCOME BASED EDUCATION

For the implementation of an outcome-based education the first requirement is to develop an outcome-based curriculum and incorporate an outcome-based assessment in the education system. By going through outcome-based assessments, evaluators will be able to evaluate whether the students have achieved the outlined standard, specific and measurable outcomes. With the proper incorporation of outcome-based education there will be a definite commitment to achieve a minimum standard for all learners without giving up at any level. At the end of the programme running with the aid of outcome-based education, a student will be able to arrive at the following outcomes:

PO1. **Engineering knowledge:** Apply the knowledge of mathematics, science, engineering fundamentals, and an engineering specialization to the solution of complex engineering problems.

PO2. **Problem analysis:** Identify, formulate, review research literature, and analyze complex engineering problems reaching substantiated conclusions using principles of mathematics, natural sciences, and engineering sciences.

PO3. **Design / development of solutions:** Design solutions for complex engineering problems and design system components or processes that meet the specified needs with appropriate consideration for the public health and safety, and the cultural, societal, and environmental considerations.

PO4. **Conduct investigations of complex problems:** Use research-based knowledge and research methods including design of experiments, analysis and interpretation of data, and synthesis of the information to provide valid conclusions.

PO5. **Modern tool usage:** Create, select, and apply appropriate techniques, resources, and modern engineering and IT tools including prediction and modeling to complex engineering activities with an understanding of the limitations.

PO6. **The engineer and society:** Apply reasoning informed by the contextual knowledge to assess societal, health, safety, legal and cultural issues, and the consequent responsibilities relevant to the professional engineering practice.

PO7. **Environment and sustainability:** Understand the impact of the professional engineering solutions in societal and environmental contexts, and demonstrate the knowledge of, and need for sustainable development.

PO8. **Ethics:** Apply ethical principles and commit to professional ethics and responsibilities and norms of the engineering practice.

PO9. **Individual and teamwork:** Function effectively as an individual, and as a member or leader in diverse teams, and in multidisciplinary settings.

PO10. **Communication:** Communicate effectively on complex engineering activities with the engineering community and with society at large, such as, being able to comprehend and write effective reports and design documentation, make effective presentations, and give and receive clear instructions.

PO11. Project management and finance: Demonstrate knowledge and understanding of the engineering and management principles and apply these to one's own work, as a member and leader in a team, to manage projects and in multidisciplinary environments.

PO12. Life-long learning: Recognize the need for and have the preparation and ability to engage in independent and life-long learning in the broadest context of technological change.

COURSE OUTCOMES

After completion of the course, the students will be able to:

CO-1: Understand the basics and breath of Civil Engineering as a field

CO-2: Understand the basics of Structural Engineering and its applications

CO-3: Understand the basics of Geotechnical Engineering and its applications

CO-4: Understand the basics of Water Resources Engineering and Sustainability and its applications

CO-5: Understand the basics of Traffic and Transportation Engineering and its applications

CO-6: Perform hands-on experiments and computations relevant to civil engineering

Course Outcomes	**Expected Mapping with Programme Outcomes** *(1- Weak Correlation; 2- Medium correlation; 3- Strong Correlation)*											
	PO-1	**PO-2**	**PO-3**	**PO-4**	**PO-5**	**PO-6**	**PO-7**	**PO-8**	**PO-9**	**PO-10**	**PO-11**	**PO-12**
CO-1	3	-	-	-	3	3	2	3	2	2	1	1
CO-2	3	2	2	1	2	-	-	-	-	-	-	-
CO-3	3	2	2	1	2	-	-	-	-	-	-	-
CO-4	3	2	2	1	2	-	3	-	-	-	-	-
CO-5	3	2	2	1	2	-	-	-	-	-	-	-
CO-6	3	3	3	2	3	-	-	-	-	3	3	-

GUIDELINES FOR TEACHERS

To implement Outcome Based Education (OBE) knowledge level and skill set of the students should be enhanced. Teachers should take a major responsibility for the proper implementation of OBE. Some of the responsibilities (not limited to) for the teachers in OBE system may be as follows:

- Within reasonable constraints, they should maneuver time to the best advantage of all students.
- They should assess the students only upon certain defined criterion without considering any other potential ineligibility to discriminate against them.
- They should try to grow the learning abilities of the students to a certain level before they leave the institute.
- They should try to ensure that all the students are equipped with quality knowledge as well as competence after they finish their education.
- They should always encourage the students to develop their ultimate performance capabilities.
- They should facilitate and encourage group work and teamwork to consolidate newer approaches.
- They should follow Bloom's taxonomy in every part of the assessment.

Bloom's Taxonomy

Level	Teacher should Check	Students should be able to	Possible Mode of Assessment
Create	Students' ability to create	Design or Create	Mini project
Evaluate	Students' ability to justify	Argue or Defend	Assignment
Analyze	Students' ability to distinguish	Differentiate or Distinguish	Project/Lab Methodology
Apply	Students' ability to use information	Operate or Demonstrate	Technical Presentation/ Demonstration
Understand	Students' ability to explain the ideas	Explain or Classify	Presentation/ Seminar
Remember	Students' ability to recall (or remember)	Define or Recall	Quiz

GUIDELINES FOR STUDENTS

Students should take equal responsibility for implementing the OBE. Some of the responsibilities (not limited to) for the students in OBE system are as follows:

- Students should be well aware of each UO before the start of a unit in each and every course.
- Students should be well aware of each CO before the start of the course.
- Students should be well aware of each PO before the start of the program.
- Students should think critically and reasonably with proper reflection and action.
- Learning of the students should be connected and integrated with practical and real-life consequences.
- Students should be aware of their competency at every level of OBE.

LIST OF FIGURES

TABLE OF CONTENTS

3. GEOTECHNICAL ENGINEERING 53-84

4. WATER RESOURCES ENGINEERING AND SUSTAINABILITY 85-110

5. TRAFFIC AND TRANSPORTATION ENGINEERING 111-154

1 Overview of Civil Engineering

1.1. Basic Understanding and History of Civil engineering

Civil engineering is the oldest discipline of engineering, and its goal is to provide people a livable and sustainable environment by combining concepts from science, mathematics, and human experiences. Bridges, dams, barrages, buildings, highways, airports, and railways are a few significant contributions of civil engineering. Apart from these major contributions, the discipline has also significantly contributed towards the development of towers, substations for electricity transmission, offshore platforms, wastewater and drinking water treatment plants, satellite space stations and traffic signals. In a nutshell, civil engineering is concerned with building the civilizational requirements of human beings.

Civil engineering covers four basic stages: planning, design, construction, and operation and maintenance. These stages can sometimes overlap as well. Planning is the first step in any civil engineering project, whether it is a government or a private project. Feasibility and environmental impact studies of the projects are also performed during the planning phase. The Design stage consists of two stages i.e., preliminary, and final. Sometimes preliminary design is also a part of the planning process. In the construction phase, physical structures of the project are erected while considering the safety and environmental concerns. The operation and maintenance phase occurs after the construction phase is completed and the project is handed over to the owner or client. The project's owner typically assumes full responsibility, although the contractor is normally obligated by a warranty agreement. The physical structures require ongoing maintenance after the guarantee period is over.

There are eight broad disciplines under the vast umbrella of Civil Engineering: Structural Engineering, Geotechnical Engineering, Environmental Engineering, Water Resources Engineering, Transportation Engineering, Construction Engineering, and Geomatics (Surveying Engineering). Structural engineering focuses on the analysis and design of the built structures. From small micro-electro-mechanical system devices to spaceships, and lofty buildings to lengthy bridges, require expertise in structural engineering. All these structures are anchored to the ground. So, there is a need to understand the capacity of the ground to bear those heavy loads. Here geotechnical engineering comes into the picture. In this branch of civil engineering, the strength of the soil and rock are investigated to provide a safe ground for structural support. This branch is concerned with the many forms of foundations that serve as a link between the structure and the ground. Environmental engineering is that branch of civil engineering in which applications of engineering principles to manage and develop water, air, and land resources for sustainable development of the natural environment is taught. Water resources engineering is a branch of civil engineering that deals with the use of water in modern society, including agricultural, industrial, household, recreational, and environmental needs. This branch also covers the occurrence and investigation of natural hazards such as floods, droughts, landslides, Glacier Lake Outburst Floods (GLOF), and flash floods like potential disasters. The field of transportation engineering is concerned with the safe and efficient transportation of people and goods. Construction engineering is a specialization that deals with the designing, planning, and management of civil engineering infrastructure. Surveying technology is newly termed as geomatics engineering and involves the application of measurement science to the collection of spatial data on land, water, and other natural or man-made objects. Geographic Information System (GIS) and satellite imagery applications are the newly added sub-disciplines of geomatics engineering.

Civil engineering works have been the bedrock of civilizational progress. From the ancient wonders to modern high-rise buildings, from drinking water supply to wastewater treatment, civil engineers have provided the building blocks for fundamental sectors such as travel, exchange of ideas, trading, and housing. It would not be an exaggeration to claim that a modern civilization would be impossible without civil engineers.

Though the oldest engineering discipline, civil engineering has kept up with the times. It provides diverse career opportunities after graduating in this discipline. With a Bachelor's degree in civil engineering, one can address challenging problems in civil engineering through government and private sectors. Government departments include the Public Works Department (PWD), Public Sector Units such as Oil and Natural Gas Corporation (ONGC), Electricity Boards, National Highway Authority of India (NHAI), armed forces, border road organization, Indian Railways, and municipal corporations. In the private sector, civil engineering graduates can join as a site engineer, assistant civil engineer, project manager, surveyors, quality-control engineer, and many more. After having post-graduation in the specialized branches of civil engineering, you can be a specialist engineer in each stream like a water resources engineer after completion of post-graduation in water resources engineering. Likewise, you could explore any branch of civil engineering and contribute to the continuous development and research needed in this industry for sustainable development.

Since prehistoric times, people have built bridges, dams, canals, amphitheaters, and roads. The building materials that are currently in use have a rich history, with some of the constructions dating back thousands of years. The background of the field of construction overlaps with that of structural engineering and a variety of other fields. Selection of material is a key factor in any type of construction. In ancient times, buildings or shelters were constructed using leaves, animal hides, and tree branches. Later materials such as mud, stone, and timber were used for durable shelters followed by bricks and cement concrete. Later, metals such as steel revolutionized the field of infrastructure construction. Nowadays, the industry makes extensive usage of the state-of-the-art composite materials such as glass, polycarbonate sheets, Un-plasticized Polyvinyl Chloride (UPVC) synthetic sheets which makes construction easier and long lasting with minimal environmental damage. The methods of construction have also evolved with time.

Stone-based construction preceded copper and bronze-based construction. Coming up to the eighteenth century, civil engineers and architects became professionals with increasing applications of mathematical models and scientific knowledge. In the nineteenth century, railways, canals, and macadam roads brought industrial revolution in the sector of civil engineering. The second industrial revolution came in the early 20th century when the use of cranes and elevators made skyscrapers and high-rise buildings possible. Computer aided designs and new technologies were also adopted in this century.

Indian Structural Marvels

Civil engineering has a long tradition in India, intertwined with the country's art and cultural history, and the country has forty UNESCO World Heritage properties, the majority of which are civil engineering marvels.

Mohenjo-Daro contains the earliest remains of Indian civil engineering. Archaeological ruins revealed that Mohenjo-Daro was a well-established town, with symmetrical and proportional buildings and roads. The buildings in Mohenjo-Daro were made of waterproof bricks, demonstrating advanced construction knowledge about three thousand years ago.

The earliest rulers of various provinces of India placed a significant importance on temples, so temples became an important exhibit of our structural history. The earliest temples in India are exquisitely designed, with intricate carvings on the walls (see Figure 1). The Kailasa temple in Ellora caves is another marvel of ancient India. It is the largest rock-cut temple which is believed to have been made from a single rock.

Figure 1: Kailasa temple in Ellora caves (Source: Wikipedia)

In medieval India, Indian structures were mostly dominated by Indo-Islamic architecture. The structures during this time showcase arches and domes. The buildings also have intricate Jaali work showcasing the importance of light in Islamic traditions. This period saw the construction of significant structures such as the Taj Mahal, Fatehpur Sikri, Red Fort, and Jama Masjid.

The British left an indelible mark on Indian architecture by introducing Indo-Saracenic style, a hybrid of Indian and European styles. Colonial architecture is visible in government buildings and residences. These structures can be found in Delhi, Bombay, Calcutta, and Shimla. Bombay University, Victoria Terminus, and Mumbai university building are examples of such architecture.

Sir Mokshagundam Visvesvaraya was a pioneer of modern civil engineering in India. He was instrumental in many civil engineering projects in India, including serving as the chief engineer of the flood protection system of Hyderabad. In India, his birthday is observed as Engineers' Day on 15th September.

1.2. Fundamentals of Civil Architecture and Town Planning

1.2.1. Civil Architecture

Architecture is the science and art of designing buildings and other physical structures. It enables us to interact with the buildings we inhabit in a way that feels both functional and aesthetically pleasing. The architects focus on functional aspects such as form and space while introducing unique characteristics that reflect the era of society to which a building belongs. Architecture not only deals with the interior and exterior of buildings but also the environment. It focuses on the larger aspects of construction, such as design of residential societies, as well as minor elements such as floor design, color theme, etc. Civil engineering and architecture play pivotal roles in the construction industry and rely on one another to accomplish the project.

Architectural Design Principles

Design, as an activity, is performed to modify the elements of the environment so that it meets the requirements. The design is not only concerned with external or internal beautification but also focuses on the function and costs of the building and involves exploring the creative and optimal solutions to a problem.

The architectural design of a building focuses on three factors: function, form, and esthetics. The design of a building primarily depends on the function of the building to be constructed, which is based on the requirements of a client and is subject to many factors such as the number of users, building bylaws, approvals, etc. After the function of a building is ascertained, the architect focuses on the form. The form of a building is dependent on many factors such as climate, economy, workers' skills, earthquake zone, etc. It also depends on the available materials, construction methods, and technologies. A residential house built in mountainous regions will have a different form (e.g., sloping roof, wooden insulation) than a house built in a plain region. The aesthetics of a building are the most apparent of the three factors and involve a lot of elements which must be carefully designed so that the building offers an inviting feeling to the users.

Architectural Design Elements

The elements are various components of the design which can be isolated and defined in such a way that each carries a unique message and contributes to the overall aesthetics of a structure. Such elements include point, line, shape, color, texture, etc. The architect utilizes these elements in the overall architectural composition to bring out the desired expression of the building. For an effective architectural composition or design, the architect follows a set of principles:

(a) Balance: Balance provided symmetry or asymmetry to the composition. It creates a state of equilibrium of weight or attention, thus conveying a feeling of stability in the structure.

(b) Harmony: Harmony is achieved when the different elements appear connected, and consistency exists in composition. If all the elements of a building are well blended, it conveys a sense of harmony among the different parts of the composition.

(c) Rhythm: The repeated use of elements like lines, shapes or colors either regularly or irregularly results in an appearance of pattern or rhythm. A rhythm could be achieved by repetition, gradation, or transition.

(d) Contrast: If an architect decides to intensify the features of opposing or dissimilar elements, the architect can do so by using contrast in the design. Contrast helps create a variety in visual appearance and can be created using elements like color, texture, or scale.

(e) Proportion: Proportion refers to the relative size of the various elements in architectural composition. A well-proportioned composition is a result of the creative use of size, color, and pattern among all the design elements and is therefore pleasant to look at.

(f) Character: The character refers to the overall style of a building that is expressed by a combination of unity, mass, proportion, etc. The character could be functional or traditional. The functional character expresses the function of the building and is often dominated by the external appearance. The traditional character is mainly governed by the culture and religious influences of the society.

A well-designed building should be functional, spacious, and aesthetically pleasant. It should exhibit the qualities of completeness, elegance, cohesion, modularity, and vision. With rising concerns about global climate change, emphasis is placed on "green building" designs that are energy-efficient and environmentally sustainable. Developing sustainable construction methods and materials that can be integrated effectively into the design and construction of buildings have become one of the most exciting problems to solve as a civil engineer.

1.2.2. Town Planning

With the growing human population, the number of industries and urban sprawl, land has become one of the most critical resources that must be managed optimally and sustainably. Town planning is a vital part of land resource management, which involves making decisions on the development and use of land. Town planning involves consideration of the current and future requirements of a community in order to develop a comprehensive development plan. It includes designing physical layouts of cities and design of buildings, roads, parks, etc., so that people find it convenient to live in. Town planning enables the government to control the growth of an urban settlement in a sustainable manner that ensures equal availability of resources and amenities to the population.

An unplanned town or city that has grown organically will become an overcrowded and unsanitary place to live in. Such unplanned cities can be seen across the world, which originated from early human settlements. The necessity of town planning arose when urban problems bloomed during the rapid industrialization in the 19th century. Without planning, the residents of a city or town would face many problems such as overcrowding, high traffic density, pollution, slums, lack of open spaces, etc. Moreover, it is nearly impossible and would incur considerable costs to correct an unplanned city.

Objectives

Generally, a city is characterized by the physical environment such as location and climate, social environment, which includes the communities dwelling in neighborhoods and the economic environment represented by the availability of business and job opportunities. The goal of town planning is to create an environment that is healthy, efficient, convenient and offers equal opportunities to the population. The primary objectives of town planning are:

(a) Avoid haphazard development of a town or city

(b) Equal availability of amenities to the population

(c) Healthy, pollution-free living environment

(d) Preservation of uniqueness and beauty of a town

(e) Convenient access to city infrastructure

As a town planner, one needs to have a creative imagination and must be able to understand the current and future needs of a city. It is required to carry out the research and prepare layout plans so that the town behaves like a living organism while carrying out day-to-day activities. Town planning is an art and science which involves exhaustive collection and analysis of information resulting in a planned physical layout that is creative, beautiful, economical, and convenient.

Process and Principles of Town Planning

Town planning is a team effort that involves various professionals such as civil engineers, architects, sociologists, geologists, administrators, and legal advisors. The planning process starts with surveying and mapping the area to identify the needs of all stakeholders, followed by the formulation of broad community goals. The objectives are identified, and possible alternative plans are developed to achieve the broad goals. The alternative plans are individually and comparatively analyzed based on various criteria, and the chosen plan is finally implemented. Overall, the process is guided by a broad set of principles:

(a) The layout of the town must not be haphazard. The housing and civil amenities should be available to all the residents.

(b) There must be open spaces, recreational centers, and reserved spaces for future development.

(c) A zoning system with rules and regulations must be strictly implemented across the town with neat grouping and distribution of public buildings.

(d) The road system must be efficient and simple, with emphasis on minimizing the commuting time from residential to work zones.

(e) The provision for a "green belt" must be made to control the town's size by having a green area around the peripheral limits.

(f) Affordable amenities such as water supply, power and health facilities must be provided to all the residents.

(g) Healthy living conditions must be maintained by preserving trees and limiting the pollution to low levels.

(h) If slums arise in the city, provisions for alternate housing and transit camps must be made to rehabilitate the slum dwellers.

1.3. Surveying

The purpose of surveying is to identify the positions of a point relative to the reference point by measuring distances, direction, and elevation. Surveying is used in every engineering project such as the construction of roads, railways, buildings, bridges, water supply, and irrigation schemes. The surveying is divided into two categories: plane surveying and geodetic surveying.

In the plane surveying, the spheroidal curvature of the Earth is ignored whereas all the lines measured in the geodetic surveying are considered as curved due to the Earth's curvature. Therefore, geodetic surveying requires a larger amount of work and has a higher degree of precision.

Classification:

The classification of the survey can be done based on the place of the survey, methods and instruments used for the survey, and the objective of the survey.

1. Based upon the place of the survey:

(a) Land Survey: This can be done in the form of a topological survey which includes linear and angular measurements to identify horizontal and vertical locations. Cadastral surveys are another type of land survey which are used for fixing property boundary lines and calculation of land area.

(b) Hydrographic Survey: This type of survey is done with the water bodies. These surveys are required for navigation purposes, harbor works, and identifying the sea level.

(c) Astronomical Survey: This survey helps in obtaining an absolute location of any point on the surface of the Earth with respect to the fixed object in space such as the sun or any fixed star.

2. Based on the purpose of the survey:

(a) Engineering Survey: This survey is conducted to provide sufficient data for designing engineering works like roads, bridges, or dam construction.

(b) Military Survey: This is used to identify the strategically critical points for military purposes.

(c) Mine Survey: This survey is done to capture the location of resources and minerals.

(d) Geological Survey: These geological surveys are used to map the different strata within the Earth's crust.

(e) Archaeological Survey: This survey is used for unearthing artifacts of ancient times.

3. Based on the instruments and methods used:

(a) Chain Survey

The chain surveying is based only on the linear measurements performed on the field. Therefore, this can be used effectively in small extent surveys of a piece of land or to take out simple details. The main station is defined as the start and end points of the chain. The subsidiary and tie station can be selected at any point of the chain. The biggest line connecting the main stations is termed as the base line. Moreover, to check the accuracy of the work conducted some *check lines* or *proof lines* are also laid. The length of the check line measured in the field shall match with the length on the plan.

The steps involved in chain steps are:

i. Reconnaissance: The whole to part method shall be employed before starting any survey. As the name implies, the surveyor should walk around the area to fix the best position for survey stations and should examine the inter visibility of survey stations.

ii. Marking and fixing survey stations: After selecting the survey stations, the marking of the stations can be done using wooden pegs if the ground is soft, nails or spikes can be used in case of roads or streets, and in case of hard grounds a cement mortar-based marker can be established.

iii. Running survey lines: After completing preliminary work, chaining can start from the baseline. The process includes chaining the line and locating the nearby objects.

(b) Tacheometric Surveying

Tacheometry uses angular surveying in which the distances of points are determined by optical means instead of other slower processes of measurement by tape or chain. However, the accuracy of the tacheometric surveying is unfavorable with that of chaining. Tacheometric is best adaptable for places having steep slopes and broken ground. The primary objective of the tacheometry is to prepare the contour maps of any region.

(c) Theodolite Survey

Theodolite is more precise equipment than chain surveying. It can measure horizontal and vertical angles, establish grades, determine differences in elevation, etc. (see Figure 2). Theodolites can be classified into two types a) Transit theodolite and b) Non-transit theodolite. The transit theodolite provides the flexibility of reversing the line of sight whereas the non-transit theodolite cannot be transited and has now become obsolete.

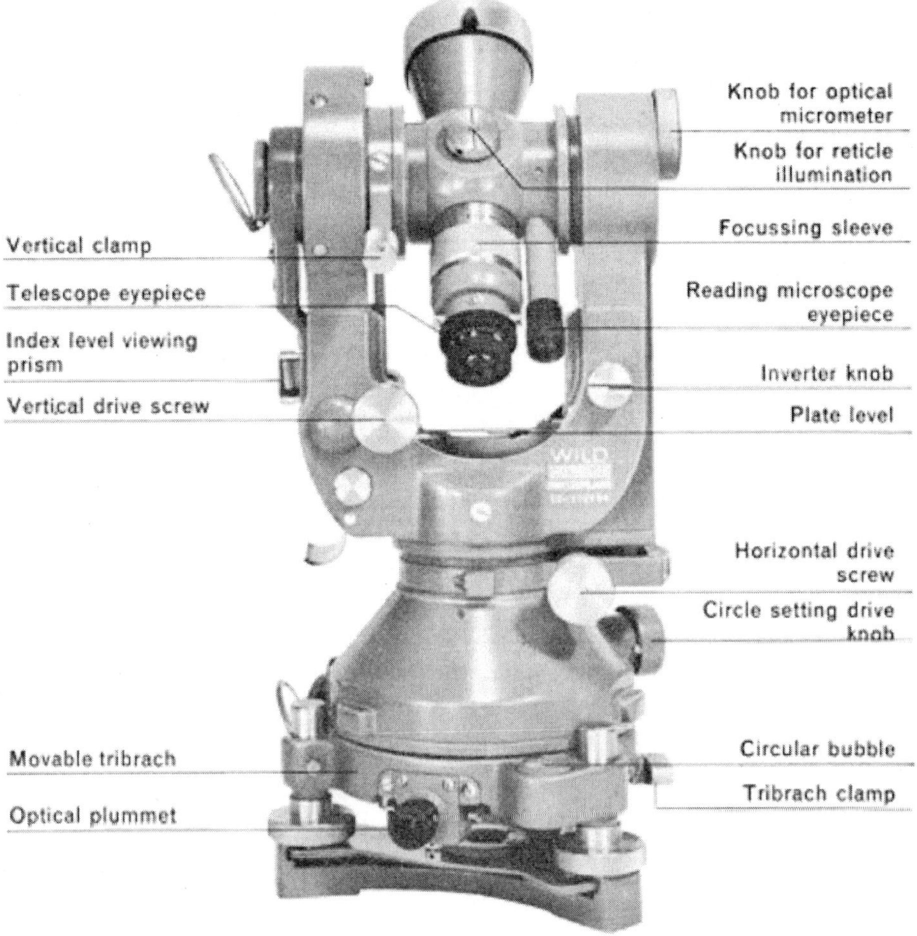

Figure 2: Theodolite in Surveying (Source: www.dreamcivil.com)

(d) Traverse Surveying

In the traverse surveying method both the direction and length of survey lines are measured. In a *closed traverse*, the survey ends at the same point from where it began, however, when the survey ends elsewhere then it is termed as an *open traverse*. Refer Figure 3. Closed traversing is used for small areas such as surveying boundaries of lakes etc., whereas open traversing is used for the long narrow strip structures such as roads, canals, or the coastline.

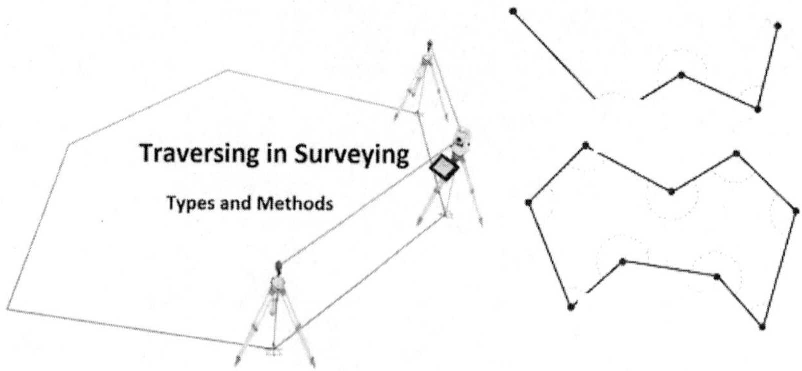

Figure 3: Traversing: a) Open Traversing; b) Closed Traversing (Source: www.theconstructor.org)

1.4. Industrial Case Studies of National Infrastructure projects

1.4.1. Bhupen Hazarika Setu

The Bhupen Hazarika Setu or Dhola-Sadiya bridge is a marvel of civil engineering. It is illustrated in Figure 4The bridge was designed and constructed despite being located in seismic zone V and facing high wind loads (design wind speed of 50 m/s) as well as strong water current (design water speed of 3 m/s). It is the longest bridge in India as of 2022 and is built over the Lohit River which is a tributary of the Brahmaputra River. The bridge was approved for construction in 2009. The construction started in November 2011 and finished in 2017. Bhupen Hazarika bridge connects the village of Dhola in the South to Sadiya in the North hence the name Dhola-Sadiya bridge. This bridge forms a permanent connection between Assam and Arunachal Pradesh and decreases the travel time by 6 hours. The Bhupen Hazarika bridge is designed to handle the weight of battle tanks making the bridge an important defense asset in the North-East.

At the location of Bhupen Hazarika bridge, Lohit River is 4100 m wide during non-flooded times. The length of the Bhupen Hazarika bridge is 9150m. The excess length is designed to handle areas adjacent to rivers subject to recurring floods, also known as the floodplain of a river. Over the floodplains of the river, the Bhupen Hazarika bridge is made using a series of arches and columns known as viaducts.

Figure 4: Bhupen Hazarika Bridge (Source: Wikipedia)

The foundation of the bridge is designed separately for river portions and floodplains. The river portion faces higher water forces, fatigue, current forces, and scour in addition to dead load from the bridge as well as the live load from vehicles. The bridge is also subject to horizontal forces from braking vehicles and seismic activities. Bhupen Hazarika bridge has 183 spans of 50m in length.

The subsoil of the lohit river consists of fine sand and for the foundation cast in situ piles have been employed. The piles transfer the superstructure load using side friction as well as tip of the pile. The structure used precast segmental concrete bridges also known as segmental PSC bridges. Each span of 50m is made up of 15 individual segments. The design of the bridge is as per Indian Roads Congress (IRC) code 112 and the loads were specified as per IRC code 6.

Seismic:

Bhupen Hazarika bridge is located in seismic zone V making it highly susceptible to earthquakes. The earthquake forces act horizontally and vertically, therefore the bridge must be designed to handle both loads. The bridge is a top-heavy structure, since the horizontal forces act through the center of gravity, the piers and piles are under high bending moment. To effectively transfer this moment from superstructure to the foundation, seismic arresters are used. Seismic arresters resist the seismic load and transfer them to the piers, therefore resisting transverse and longitudinal forces. The bridge is designed with an adequate factor of safety to handle vertical seismic loads.

Impact of flash floods:

Flash floods are high-intensity floods of small durations. The flash floods in the Lohit River can be caused by rains as well as lake outbursts, which implied that the construction of the Bhupen Hazarika bridge was susceptible to flash floods during the rainy as well as the dry seasons. To counter the impact of flash floods, lookout posts upstream of the lohit river were set up which provided an early warning in flash floods. Once early warning was received the equipment and manpower were removed out of the way until the flash flood passed. Despite these measures, a lot of equipment was lost to the flash floods during construction.

1.4.2. Delhi Metro

The urbanization of India is happening at an unprecedented pace. However, most cities rely on road transport, which leads to the congestion of the roads, which is a big problem for cities. Public transport is the most viable method to improve a city's transportation system. In India, most cities do not have adequate and efficient public transport. Delhi is one of the cities facing the same problem in the case of road transportation. Delhi met high population growth from 1961 to 1981, from 2.6 to 6.2 million people, which increased to 13.7 million by 2001. In addition, from 1981 to 1995, vehicle registration also increased (by 329%) to 2.5 million leading to greater traffic congestion and air pollution. These challenges not only waste time and resources but also affect public health. A traffic and travel characteristics study conducted in the city in 1969 stated the requirement of a mass rapid transit system (MRTS) for New Delhi. The Urban Arts Commission proposed establishing a multi-modal transportation system in 1984. To tackle this public transportation problem, the Government of India (GoI) and the Government of National Capital Territory of Delhi (GNCTD) formed a joint venture known as Delhi Metro Rail Corporation Ltd. (DMRC) in 1995 under the Companies Act 1956 with equal equity to provide a passenger-friendly, eco-friendly, and world-class mass rapid transport system.

The first challenge of the DMRC is to raise funds and appoint the managing director. The project's funding was provided by the Indian government and the Japanese International Cooperation Agency (JICA), and Mr. Elattuvalapil Sreedharan was appointed as the first CEO of the DMRC. The construction of the Delhi metro started in 1998 and was divided into four phases (1995-2006; 2006-2011; 2011-2016; 2016-2021). The Delhi Metro has several advantages, including reduced air pollution, passenger travel time savings, reduction in number of accidents, road traffic reduction, and fuel savings. DMRC became the first-ever railway project certified by the United Nations (UN) to get carbon credits for reducing greenhouse gasses by 6.3 lack tons annually in the world.

The metro construction in old cities like Delhi raises issues about whether it should be elevated or underground. The government of India and DMRC both opposed metro underground construction except for technical issues and near the historical monuments. They also highlighted various safety and financial matters. According to the DMRC's experience, underground construction cost is more than two times the elevated construction cost. Although building an underground metro is expensive, the Delhi metro structure is both elevated and underground. Due to the aesthetic view of the historic buildings, one cannot provide the elevated metro line near the historical monuments. The tunneling is most challenging in old Delhi areas as the foundation of old buildings is weak, so most tunnels are 20 m below the ground label. DMRC used a Tunnel boring Machine (TBM), an alternative to the drilling and blasting (D&B) method for urban areas, as it limits the disturbance in surrounding areas and reduces the cost of lining the tunnel (Figure 5). The initial two phases of the Delhi metro were constructed in a broad-gauge system; after that, it was changed to a standard gauge because over the world, metro lines have a standard gauge that can help in the technology transfer. Delhi metro used basalt-less-tracks for elevated and underground tracks, which required less maintenance than tracks with basalt, making it economical.

Figure 5: TERRATEC EPB TBM used for tunneling in the Delhi Metro (Source: https://tunnelingonline.com/)

Ridership in the Delhi metro is continuously increasing. According to the CAG's report, since operating in 2002, average ridership was eight lakh passengers per day, which increased to 27.76 lakh passengers per day in the financial year 2019-2020. Moreover, it reduced the congestion in road transport. It was found that between 2004 and 2006, the Delhi metro reduced CO pollution by 34% and NO2 by 31% at the ITO metro station, along with significant reduction in the PM2.5 particles.

Currently, the Delhi metro has a 391 km long rail network with 286 stations and 12 lines, which is India's largest metro rail network. The metro project reduces travel time and provides safe and economical transportation. This project is one of its kind as it was built in one of the most crowded cities in the world with significant engineering challenges. Now DMRC is working as a project consultant in many metro projects in India as well as foreign projects such as Jakarta mass rapid transit project, Kuwait City.

1.4.3. Atal Tunnel

The Atal Tunnel is a highway tunnel in Himachal Pradesh that connects the Kullu valley with the Lahaul and Spiti valleys. It is located beneath the Rohtang Pass. The tunnel is 9.02 kilometers long and is the longest above 10,000 feet. The tunnel was developed by the Border Roads Organization (BRO). Rohtang in Ladakhi means the 'Pile of corpses' due to the thousands of people who had died over the centuries when they crossed the pass in winter. The Rohtang pass receives heavy snowfall during winters and remains closed for 6 months thereby cutting all road connectivity to Lahaul valley. Due to sudden snowfall every year vehicular traffic gets stuck in

Rohtang pass which then takes days to clear up. Atal tunnel also provides a very important alternative to Srinagar-Kargil route for Leh Ladakh.

The Atal tunnel has provided the following major benefits:

(a) Year around connectivity to Lahaul Spiti, Pangi, and Ladakh region

(b) Increased amount of tourism leading to overall development of the area

(c) Reduced the recurring expense of maintaining of Rohtang pass

(d) Reduced the cost of airlifting supplies

The earliest preliminary survey, pre-feasibility study, and geological investigations were done by Geological Survey of India (GSI) in 1984. This study formed the basis for investigation of further studies. The preliminary report covered the tunnel alignment, geological aspects, and tunnel alignment. A detailed preliminary report was prepared by BRO highlighting the geological conditions, tunnel alignment, construction, ventilation, lighting, safety, and cost.

The Detailed Project Report (DPR) included surface studies, tunnel alignment and laboratory tests, techno-economic studies, designs, construction planning and equipment, leakage studies, dewatering, grouting, blasting and cost studies. The studies were carried out using Remote Sensing, Geological photos, Differential Global Positioning System (DGPS) etc.

The Atal tunnel has two ends. The North end is located across Rohtang pass on the Southwest bank of Chandra River. The excavation of tunnels was done from both the ends. Since the North end is covered in snow during winters, three-fourth of excavation was done from the South end. The tunnel is a two-lane bidirectional tunnel with horseshoe shape and was excavated using New Austrian Tunneling Methodology (NATM).

Based on the cost of construction and the stress on the tunnel, the tunnel features a single tube with two lanes and a horseshoe-shaped cross section. On both sides of the main tunnel, there are 4 m wide roads and 1 m wide footpaths, with a 0.5 m median separator in the center, and the overhead clearance is 5.525m. The final concrete lining is 50 cm thick. The emergency egress tunnel which is used in case of an emergency is not in parallel to the tunnel but is located under the main tunnel.

The support system was designed based on the characteristics of the rock and the depth of the overburden. Numerical modeling was used to develop the detailed design of the support system based on data obtained from drilled cores. The support consists of shotcrete, steel fiber reinforced shotcrete or wire meshes. The rock bolting was done using self-drilling rock bolts. The final lining was cast-in-situ plain concrete, with the exception of the Seri Nala portion, which had reinforced concrete.

The capacity of fresh air required in the tunnel is decided based on vehicular traffic, vehicular emissions, and turbidity limit inside the tunnel. Therefore, a semi transverse ventilation system was used and ventilation fans which insert fresh air were installed in the ceiling of the tunnel. On 15th October 2017, the tunnel's final breakthrough was made thereby connecting both sides of the tunnel. With the breakthrough the North portal became accessible for the year around since then supply can be provided from the south portal during winters. Also, due to natural ventilation, the temperature and natural airflow were established.

Problems faced while construction:

1. The biggest problem was the Seri nala fault zone which took 4 years to overcome. Due to weathering caused by the Seri nala only muck and water started flowing. The water was flowing at a rate of 8000 liters per minute. This was not anticipated in any of the surveys as well as DPRs. Finally, the DRESS (Drainage Reinforced Excavation and Support Solutions) tunneling method was employed, the muck was stopped with sandbags, and a thick layer of shotcreting was applied. Drainage holes were also installed every 3-4 meters to channel water.

2. The tunnel's North side was composed of hard, brittle rocks like gneisses, which were prone to bursting. The rock load increased as a result of the delayed stress release, causing cracks in the installed lining. Even the rock bolt plates were bent due to the high stress. The installation of lining stress controllers solved the problem. This problem was remedied by installing lining stress controllers.

3. The South end of the tunnel, where the majority of the work was supposed to be done, was in a landslide zone. This posed a significant challenge because the ventilation building, and control room are all located on the southern end of the tunnel. To remedy the impact all the important structures were relocated.

4. The Atal tunnel is located at a height of 3100m in the Himalayas where the temperature gets very low. The workers were susceptible to hypothermia, frost bites etc. Therefore, the workers were provided with special clothing and kits. Not only limited to high altitude sickness, but the Atal tunnel was also being built when Covid-19 happened which caused temporary cessation of work.

5. The tunnel was constructed through a crystalline thrust located in the young Himalayas which added to the complexity of construction.

6. When compared to other tunnels around the world, the Atal tunnel has a very high overburden. The highest overburden is almost 2kms.

Salient features of the tunnel:

1. In case of a mishap the tunnel has a one-of-a-kind escape configuration in which the escape tunnel is inside the main tube itself. There are multiple escape cuddies located at every 500m.

2. The tunnel has generator backup at each portal. The tunnel gets a supply of 11 KVA from the south end and a supply of 33 KVA is scheduled from the north.

3. The tunnel has semi-transverse ventilation, which is powered by three axial fans with a combined capacity of 630 kW, it is the most powerful tunnel ventilation in the country.

4. The electrical and mechanical systems are well suited to operate under extreme conditions(-40°C).

5. Heat tracing units are installed in fire hydrant pipes to prevent water from freezing, which is necessary in the event of a fire.

1.4.4. Namami Gange

The Ganga River, originating in the Himalayas and flowing to the Bay of Bengal covers more than 2,500 km in northern and eastern India (Figure 6). The Namami Gange project was initiated with the intention of conserving, reducing pollution, and revitalizing the Ganga River. The project was approved with a budget estimate of Rs. 20000 crores by the government in June 2014.

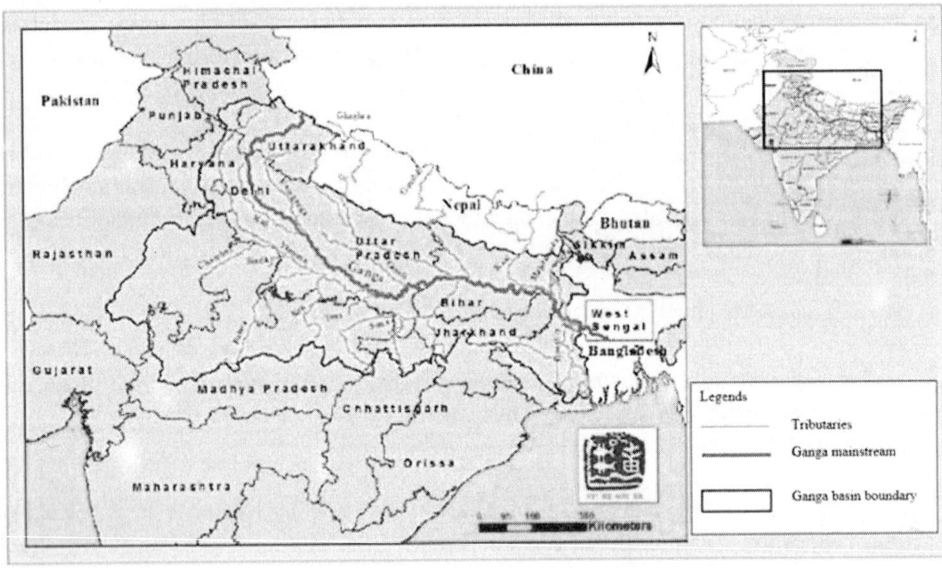

Figure 6: Namami Gange Project Details

Need of the Project

- The Ganga basin covers approximately 26% landmass of India which further extends to the neighboring countries including Nepal, China, and Bangladesh.

- The river Ganga contributes to economic development, impacts the environment, and holds significant cultural importance in India.

Objectives of the Project:

•To establish systems in place such that individuals, institutions, and corporate organizations can effectively contribute to the rejuvenation of the river Ganga.

•To promote technological innovation and facilitation of such interventions into the ground reality.

•To encourage local community participation by bringing mass awareness regarding the connection of the rivers to the lives of the common public.

Key Plan: The Ganga River gets polluted by the extracts of industry wastage, municipal wastewater, solid waste, and other diverse sources such as partially cremated bodies. The focus of this project was to ensure the continuous flow of the unpolluted water in the river Ganga. The plan of the project was prepared using the following eight key pillars into consideration.

(a) Sewerage Treatment Infrastructure

(b) River-Surface Cleaning

(c) Afforestation

(d) Industrial Effluent Monitoring

(e) River-Front Development

(f) Biodiversity

(g) Public Awareness

(h) Ganga Gram

The project was implemented by establishing multilevel monitoring systems at the national, state, and district levels.

1.4.5. Mumbai-Delhi Expressway

This is an 8-lane (expandable to 12-lane) expressway with a length of approximately 1,350 km, connecting Haryana, Rajasthan, Madhya Pradesh, Gujrat, and Maharashtra (Figure 7). This expressway is economically important as it connects the national capital Delhi to the financial capital of the country. This is an access-controlled expressway having roadside amenities at an interval of approximately 50 km.

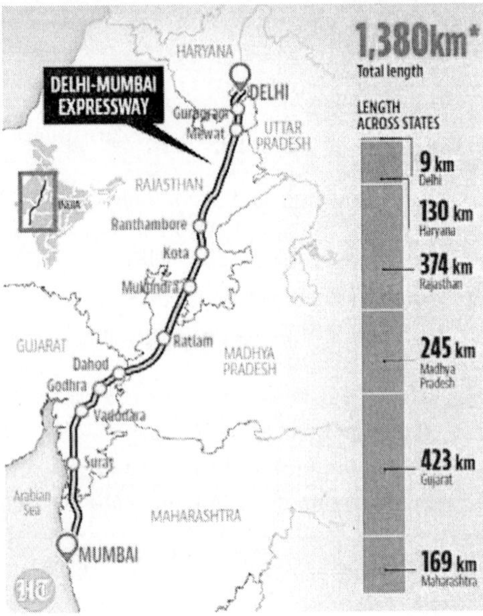

Figure 7: India's longest expressway on track to bring two metropolises closer (Source: Hindustan Times)

Objectives of the Project

The primary objective of the transport facility is to ease the commute between origin and destination. This objective can be achieved by reducing travel cost, and travel time, and increasing the comfort and safety of the passenger and goods. Mumbai and Delhi are two very important strategic locations for India and the alignment of this expressway is such that the distance between Mumbai and Delhi reduces by about 150 km and travel time to 13 hours from 24 hours. The expressway is designed to provide better ride quality, safety, and convenience due to roadside amenities.

Special Features:

(a) Wayside amenities: The expressway is designed considering the comfort of the commuters. The wayside amenities include ATM, retail shops, food shops, and hotels. It also has charging stations for the electric vehicle along with several fuel stations. The roadside amenities also accommodate the trauma centers for addressing the road crash victims immediately.

(b) Environment friendly: The expressway is surrounded by over 20 lakh trees. The facilities of drip irrigation along with a rainwater harvesting system have been designed at every 500 meters.

(c) Wildlife crossings: To consider the safety of both the high-speed drivers and wildlife animals, five natural-looking wildlife crossings are provided between tiger reserves. The crossing corridor has 8 meters tall side barrier walls on both sides.

1.4.6. Sardar-Sarovar Project

The Sardar Sarovar project serves water and electricity to four states of India including Rajasthan, Gujarat, Madhya Pradesh, and Maharashtra (Figure 8). This is the nation's largest water resource project with discharge capacity of 30.7 cubic meters per second. The Sardar-Sarovar dam is 1210 meters long and 163 meters in height. This is the third-highest dam in India after the Bhakra (226 meters) in Himachal Pradesh and Lakhwar (192 meters) in Uttarakhand. This is a gravity dam involving 6.82 million cubic meters of concrete.

Figure 8: Sardar Sarovar Project: the largest water resources projects in India

Objectives of the Project

(a) Irrigation: The construction of this dam facilitates the irrigation of approximately 18,450 million m^2 of land which includes 3112 villages of Gujarat, 2,460 million m^2 of land covering deserted districts of Rajasthan, and 375 million m^2 area of Maharashtra. The major areas covered through this dam are the drought-prone areas of Rajasthan and Gujarat.

(b) Drinking water supply: The dam provides drinking water to 9490 villages serving approximately 28 million population of Gujarat. The demand of several industries for water is also fulfilled by the dam, increasing the income opportunity and economic growth of the country.

(c) Power**:** The dam also facilitates power generation through two powerhouses with the capacity of 1200 Megawatt (MW), and 250 MW. Madhya Pradesh, Maharashtra, and Gujarat share the generated power.

(d) Flood Protection: This dam also acts as protection from floods covering a region of 300 million m^2 protecting about four lakh people in Gujarat.

1.5. Professional ethics

How relevant is ethics for a civil engineering student? To highlight its importance, we are first going to understand what ethics are and what is the relevance of ethics in the civil engineering profession. The fundamental responsibility of an engineer is to secure the health, safety, and well-being of the general public. The ultimate recipients of engineering services are the general public and hence the ethical concerns that arise during the decision-making problem are of utmost importance. At every step of engineering, from acquiring the materials to the actual completion and even after sales, ethics becomes really important. However, establishing a professional career as an ethical person does not guarantee that you will be able to solve complex problems. Individuals who are aware of professional ethics will better be able to determine what actions are appropriate for a given problem. The purpose of this section is to give an overview of ethics in engineering and to describe how to resolve ethical dilemmas using standard ethical codes.

In the book Ethics in Engineering by Martin and Schinzinger, they explain ethics as "*synonymous with morality, it refers to moral ideals that are allowed (all right), regulations and laws that are desirable.*" In a perfect society, there would be no need for ethical compliance because everyone would strive to be the best possible version of themselves. However, we do not live in an ideal world.

Professional ethics is a set of rules that regulates the behavior of a person or group in the workplace. It is difficult to distinguish between professional and personal ethics because in both cases, people make decisions. When it is on an organizational, stakeholder, or public level, it is in the first group. When it is about how one person interacts with others in everyday life, it is in the second group. Ethics is not a law in the legal sense, but they govern what is morally proper and acceptable in the engineering profession. Individual integrity and honesty, which civil engineering students should learn and execute as early as possible in their education, are intricately linked. Because of its importance in society, ethics is admired and held in high regard by others.

Most of the civil engineering curriculum is spent on technical subjects, and their professional expertise is focused to implementing these technical skills to solve society's problems. Rather than an accumulation of information and power, professionalism is a technique of thinking and living. In resolving new challenges, the professionals must be able to incorporate previously encountered experiences and ideas. Such problem solving necessitates independent thinking and non-self-serving motivation. An engineering professional's improper behavior or remark is equally damaging whether it is made at work or on vacation when one acquires the commitment associated with specific engineering training and a professional job, one also acquires the authority to behave within a set of norms and actions that will not impair the trust that society thrusts upon the engineering profession. Hence, professionals are expected to perform their duties in accordance with high ethical standards and with constant regard for the public good.

A professional's ethical maturity is significant since it indicates how the individual tackles a value conflict. Professionals have responsibilities to their organization, client, profession, society, and themselves. Professionals also have environmental and sustainability obligations. When making a decision that contains conflicting values, an ethically mature person realizes the relevance of all these value commitments and properly weighs each of them.

A code of ethics is a set of rules which helps to determine what is right and wrong. When you are faced with a dilemma and must decide without knowing how to continue, a code of ethics gives a foundation or guideline for your ethical decision-making. Each professional organization in engineering has a code of ethics that specifies their obligations. Engineers are accountable for the environment as well as public health, safety, and welfare. In addition, they are expected to avoid conflicts of interest, be honest, maintain professional competence, not engage in unfair competition, and make only unbiased decisions and true remarks. A code of ethics is a key set of principles and behaviors designed to be followed by users so that they behave responsibly, in an unimpeachable manner, with honor and integrity.

Code of ethics by American Society of Civil Engineer's (ASCE) is one of the most widely known codes for civil engineers. It is targeted at three groups: the public, employers and clients, and licensed civil engineers. Figure 9 explains the fundamental principle of code of ethics by ASCE. Engineers retain and enhance the honor, integrity, and dignity of their profession by keeping the 4 principles explained in the figure.

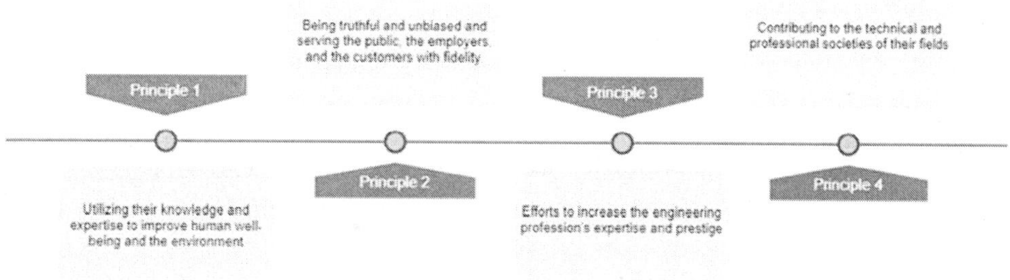

Figure 9: ASCE's Fundamental Principles

The Engineering Council of India's code of ethics is one of the most well-known codes of ethics in India. According to them, engineers apply scientific knowledge to solve the problem. The quality of life and all parts of the economy are affected by professional engineers and consulting engineering firms. Hence ethics are crucial to the profession's values. Engineers should follow this code of ethics when working with the public, clients, employers, workers, and other engineers. A total of three main points is presented, each with numerous sub points (Figure 10).

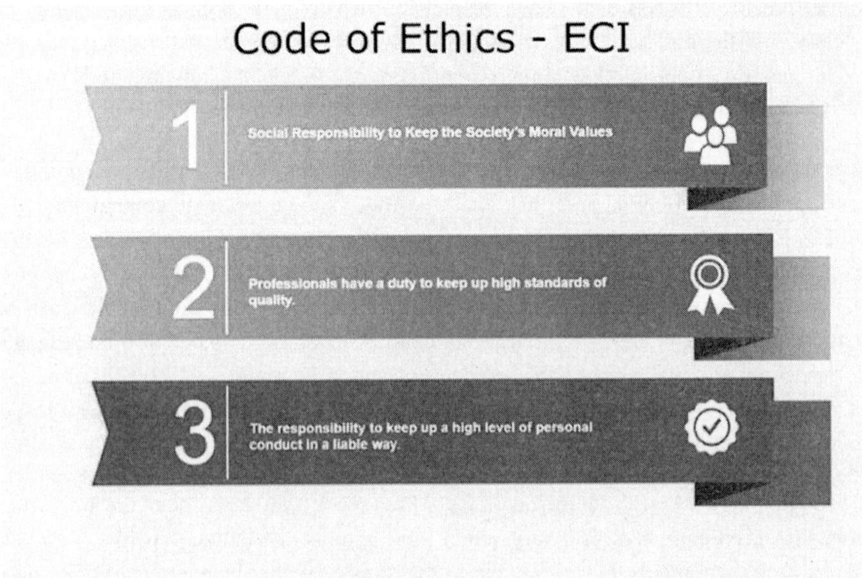

Figure 10: The three articles provided by the Engineering Council of India in their code of ethics

It is essential to discuss the rights and responsibilities of engineers in relation to their professional responsibilities. Responsibility is a duty or obligation to properly accomplish or complete a task imposed by someone or generated by one's own promise or circumstances, which bears a consequence for failing. The rights and responsibilities of an engineer includes maintaining workplace quality, ensure employees and public safety, guarantee of legal compliance, avoid conflicts of interest, monitor outside employment/Activities that won't conflict the particular organization, preserve confidential or proprietary information, maintaining accurate and complete records, use and protection of employers assets, gifts, meals, services and entertainment, Relationship with competitors, avoid Bribes and kickbacks, follow guidelines in purchases of goods and services, environmental protection, whistle blowing if needed, relationships with clients, outside contractors, and Consultants.

An ethical climate is a mix of an organization's formal procedures, informal traditions, and practices, and the personal attitudes and commitments of engineers and other stakeholders. For a better society it is essential to promote an ethical climate both organizationally and individually. Leadership on value, professional, and ethical issues will always be required, regardless of the size or type of the organization. Management and professional requirements may fluctuate depending on the size and work environment, yet leadership on value, professional and ethical concerns will always be important.

In the next few chapters, details of various subfields of civil engineering such as structural engineering, water resources engineering, transportation engineering, etc. are discussed.

2 Structural Engineering

Unit specifics

Through this unit we have discussed the following aspects:

- *Introduction to mechanics of structures*
- *Brief introduction to types of structures*
- *Fundamentals of Building Materials*
- *A brief discussion on sustainability on construction*
- *Introduction to construction and contracts management*
- *A brief discussion on repairs and rehabilitation of structures*

The practical applications of the topics are discussed for generating further curiosity and creativity as well as improving problem-solving capacity.

Besides giving a large number of multiple-choice questions as well as questions of short and long answer types marked in two categories following lower and higher order of Bloom's taxonomy, assignments through a number of numerical problems, a list of references, and suggested readings are given in the unit so that one can go through them for practice. It is important to note that for getting more information on various topics of interest some QR codes have been provided in different sections which can be scanned for relevant supportive knowledge.

After the related practical, based on the content, there is a "Know More" section. This section has been carefully designed so that the supplementary information provided in this part becomes beneficial for the users of the book. This section mainly highlights the initial activity, examples of some interesting facts, analogy, history of the development of the subject focusing the salient observations and finding, timelines starting from the development of the concerned topics up to the recent time, applications of the subject matter for our day-to-day real life or/and industrial applications on variety of aspects, case study related to environmental, sustainability, social and ethical issues whichever applicable, and finally inquisitiveness and curiosity topics of the unit.

Rationale

This fundamental unit on structural engineering will help students to get a basic idea about primary elements and concepts of structural engineering. The chapter starts with an introduction to the fundamentals of structural mechanics explaining the basics of fundamental concepts and elements that are relevant to structural mechanics. A brief discussion on the types of civil engineering structures is put forward in the next section. This chapter then explains the various types of building materials and their properties. This chapter also discusses the need for sustainability in construction practices. The basics of contract and construction management are also discussed in this chapter. This chapter lastly gives a brief discussion on the repair and rehabilitation of civil engineering structures. The topics covered in this chapter aim to develop the basic idea of structural engineering. Various examples are mentioned in the chapter to impart a basic and fundamental understanding of the subject in the field of civil engineering.

Structural engineering is the branch of civil engineering that deals with the design and construction of civil engineering structures. A large part of the spectrum of civil engineering is covered by structural engineering.

Pre-requisites

Mathematics: Coordinate Systems (Class XII)

Physics: Mechanics (Class XII)

Unit outcomes

List of outcomes of this unit is as follows:

U1-O1: Understand the basics of mechanics of structures.

U1-O2: Get an idea about the different types of structures.

U1-O3: Get an idea about the various building materials used in civil engineering.

U1-O4: Understand the importance of sustainability in construction.

U1-O5: Understand the basic idea of construction and contract management.

U1-06: Get an idea about the repair and rehabilitation of structures.

Unit-1 Outcomes	***EXPECTED MAPPING WITH COURSE OUTCOMES*** *(1- Weak Correlation; 2- Medium correlation; 3- Strong Correlation)*

	CO -1	CO -2	CO -3	CO -4	CO -5	CO -6
U1-O1	3	3	3	-	3	1
U1-O2	1	1	2	2	1	-
U1-O3	2	1	3	1	2	1
U1-O4	-	-	3	1	2	2
U1-O5	3	3	3	-	3	1

Structural engineering is the discipline of civil engineering concerned with the design and construction of civil engineering structures such as buildings, water-retaining structures, bridges, frames, trusses, and retaining walls. A structural engineer considers stability, strength, and stiffness of a structure while designing so that it can withstand the forces in an economical and safe manner. Structural engineering applies the concepts of mechanics in designing various engineering structures. It deals with the design, building, and maintenance of the infrastructure using mathematical equations.

2.1. Introduction to Mechanics of Structure

Mechanics is the study of the state of a body under the application of external forces and energy. The body can be rigid or deformable. If a body does not bend, twist, or change its shape on the application of force, then the body is called a rigid body. It is an ideal condition as no material is rigid. Even diamonds, the hardest material on earth, change their shape when exposed to extreme force. Whereas if the body bends, twists, or changes its shape by applying any type of force, then the body is called a non-rigid or deformable body.

A structure is defined as a system of interconnected members assembled in such a way that it can withstand load under stable conditions. Mechanics of structure studies the effect of the application of external force on a non-rigid structure and helps us to understand the behavior of the structure because of the application of the force. It studies how the external forces acting on the structure would result in translational, twisting, or rotational effects on the body or a combination of two or more effects. In practice, a structure is subjected to various artificial loads like loads coming from machinery, vehicles, household objects, etc., and natural loads like earthquakes, wind, etc. Hence, understanding the behavior of the structure subjected to various external loads is necessary for the stable and safe design of a structure.

2.1.1. Basic concepts in structural mechanics

1. Space

Space is the geometric region occupied by bodies whose positions are described by linear and angular measurements relative to a coordinate system. For example, for 3-dimensional problems, three independent coordinates are required.

2. Time

Time is the measure of the succession of events and is a basic quantity in dynamics. Time is not directly involved in the analysis of a statics problem.

3. Mass

Mass is the measure of the inertia of a body, which is its resistance to a change of state of velocity. Mass can also be thought of as the quantity of matter in the body. The mass of a body affects its gravitational attraction force with other bodies. This force appears in many applications in statics.

4. Force

Force is the action of one body on another and tends to move a body in the direction of its action. The action of a force is characterized by its magnitude, the direction of its action, and its point of application. Thus, force is a vector quantity, and it is described in detail in the next chapter.

5. Particle

A particle is a body of negligible dimensions. In the mathematical sense, a particle is a body whose dimensions are near zero so that it can be analyzed as a mass-concentrated point.

6. Rigid body

A body is considered rigid when the change in distance between any two points is negligible for the purpose at hand. For instance, the calculation of the tension in the cable which supports the boom of a mobile crane under load is unaffected by the small internal deformations in the structural members of the boom. For the purpose, then, of determining the external forces which act on the boom we may treat it as a rigid body in equilibrium. Determination of internal deformations belongs to the study of the mechanics of deformable bodies, which normally follows statics in curriculum

7. Scalars and vectors

Scalars are quantities with magnitudes and no direction and vectors are quantities that have both magnitude and direction. For example, pressure is scalar because it does not take into consideration any direction associated with it whereas force is a vector because direction along with its magnitude is relevant while considering a force. A force acting on an object of the same magnitude and different direction will have a different impact on the object.

8. Equilibrium

When all the forces and moments acting on a body are zero, then the body is said to be in equilibrium. In plane equilibrium (2-dimensional) conditions (for bodies like trusses, beams, etc.) a body is in equilibrium if,

$$\Sigma Fx = 0, \Sigma Fy = 0, \Sigma Mz = 0.$$

where F is force, M is the moment and x, y, z are the 3 dimensions

9. Center of Gravity

The center of gravity (CG) of a body is the hypothetical point through which the force of gravity acts on an object. Any external force acting through the center of gravity cannot produce any rotational effect on the body. Hence, the center of gravity is of concern in structural engineering for the design of a stable structure. It is attempted to keep the center of gravity as close to the ground surface as possible in the design of structures to get better structural stability.

10. Stress

Stress is a physical quantity that expresses the internal force exerted by the neighboring particles of the continuous material. Stress is defined as the force applied per unit area and is the reaction of the structural members to its self-weight and the external loads applied to it. For example, when a solid vertical bar is supporting an overhead weight, each particle in the bar pushes on the particles immediately below it. Stress is frequently represented by the Greek Letter Sigma (σ). Depending upon the nature of force applied, stress can be classified into five types as tensile stress, compressive stress, shear stress, torsion, and combined stress.

- Tensile stress: Tensile stress is a physical parameter that is associated with stretching and tensile force. It is equal to the tensile force per unit area of the surface of the body on which it acts. It occurs when the material is stretched along the axis to which the load is applied (Figure 11).

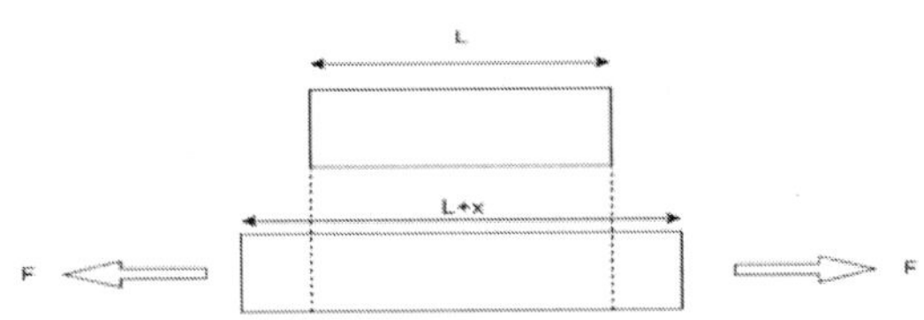

Figure 11: Tensile Strength of Material.

Compressive stress: Compressive stress is a physical parameter that is associated with compression. It is equal to the tensile force per unit area of the surface of the body on which it acts. It occurs when the material is compressed along the axis to which load is applied (Figure 12).

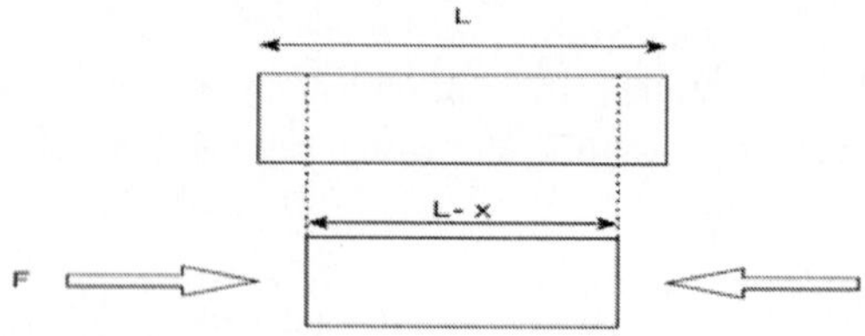

Figure 12: Compressive Strength of Material.

•Shear stress: When a force is applied causing a sliding effect, stress is produced between two surfaces, and this stress is known as shear stress. It is equal to the force per unit area of the sliding surface of the body (Figure 13).

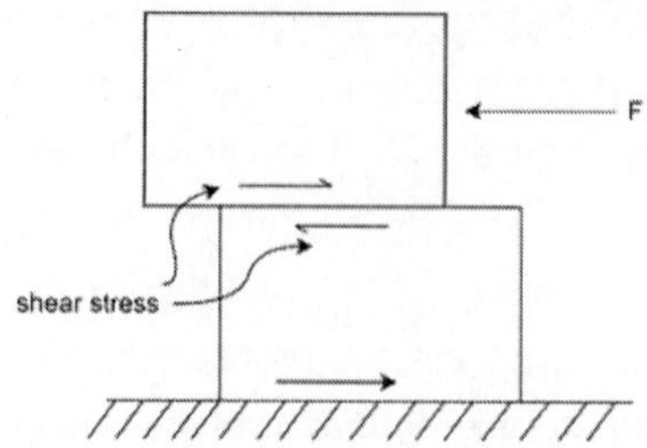

Figure 13: Shear stress between two surfaces

•Torsion: - Torsional stress can be defined as the shear stress that acts on a transverse cross-section which is caused by the action of rotation or twist (Figure 14).

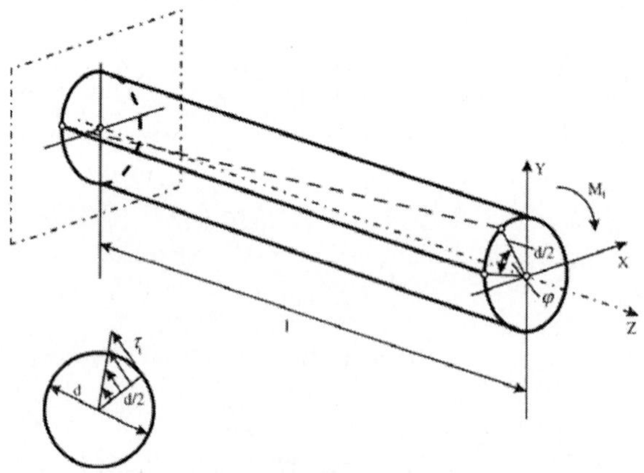

Figure 14: Torsion in a circular rod. (Source: Ceschinski, H. et al., 2000).

•Combined stresses: - When two or more stresses act simultaneously on any structure, the combination of these stresses is called combined stresses. There are various combinations of stresses such as bending and shear at the same time, twisted and stressed at the same time, or twisted and compressed at the same time.

11. Bending moment

When an external load is applied to any object, moments are generated causing a bending effect on the member. The net resultant moment due to external loads and moments that cause bending at a point is called the bending moment at that point. The bending moment at any point in a member due to a load is equal to the load multiplied by the perpendicular distance of the point where the bending moment is to be calculated from the point of application of the load.

$$Bending\ moment(BM) = Force(F)\ \times Perpendicular\ distance(L)$$

Figure 15 shows a cantilever beam. The bending moment at point A due to load P acting at a distance of x from A is equal to P_x.

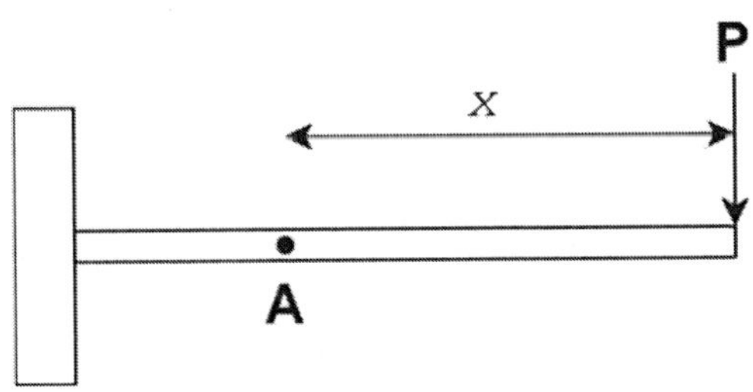

Figure 15: Load diagram leading to bending moment

2.1.2. Load

The force exerted on any surface is termed as load. A load may be applied in several directions and based on the direction of load, can be categorized into vertical loads, horizontal loads, and special loads, which can be further classified as well.

1. Vertical load: The vertical load consists of live load, dead load, and snow load which acts in the direction of gravity or opposite to it.

a. Live load: The load which can move from one place to another is called live load. Live loads are dynamic loads like furniture, water, people, and equipment whose position can be easily changed. It changes from time to time depending on the no. of people present in the structure.

b. Dead load: Dead load refers to permanent and stationary loads. Dead loads are static loads such as the weight of a structure, machine, equipment, or material. Steel, brick, sand, wires, cement, and other materials that are not frequently replaced and remain in place for an extended period are examples of dead loads.

c. Snow load: When snow accumulates on the roof during the winter season, it exerts a load on the structure, which is referred to as snow load. The amount of snow that accumulates depends on the type of roof and the area.

2. Horizontal load: The horizontal load consists of wind load and earthquake load.

a. Wind load: The force exerted by the wind along the elevation of a building is called wind load. It is the most powerful force operating on a tall building. It changes depending on the size, elevation, and style of the structure. Some of the structures are taller, but the force on them is smaller. Some structures are short, yet the wind force bearing on them is significant. It is because of the way the building is designed. The wind load varies according to the geographical area. Wind generates three kinds of loads on the structure: uplift, shear, and lateral.

b. Earthquake load: It is caused by a seismic action of the soil, which causes inertial forces to impact on the structure of the building. A fast-loading motion damages the building. Earthquake loading is proportional to building mass. The greater the mass, the greater the possibility of the building falling.

3. Special load: The special load consists of erection load, soil load, and flood load.

a. Erection load: The load exerted on a structure because of the rigging and derigging of any substance, structural, or other material is known as the erection load. It is a transient load acting on the structure.

b. Soil load: Soil load acts on structures where one side of the structure is filled with soil. Soil load is observed in soil retaining structures or underground structures.

c. Flood load: Flooding puts pressure on buildings and infrastructure. It is a natural burden that occurs quite rarely in building life.

2.2. Types of structures

A building is a structure with walls and a roof situated in one location more or less permanently. Buildings occur in a broad range of sizes, forms, and applications, and they have been modified over history for a range of causes, including the availability of construction materials, weather, land costs, ground conditions, particular uses, prestige, and aesthetic considerations. Home construction may be clearly seen dating back to around 18,000 BC. During the Neolithic age, buildings spread. There are various types of structures like buildings, tall structures, bridges, water retaining structures, wind tunnels, Power plant structures, etc.

2.2.1. Buildings

Buildings may be divided into the following categories according to the nature of their usage or occupants

(a) Residential building: A building in which people live and do their daily activity is called a residential building. For example, Flats, villas, lodges, houses, apartments, hotels, hostels, etc.

(b) Educational Building: Buildings that are used for educational purposes such as colleges, schools, libraries, etc.

(c) Assembly buildings: Buildings that are used for religious festivals and for entertainment. For example, Temple, Church, Cinema, etc.

(d) Institutional building: Building which is used for health, mental, physical, and medical institutions are known as Institutional Buildings. Example: Hospital, Sanatoria, etc.

(e) Business building: Buildings used for doing any type of business such as banks, shops, etc. are called business buildings.

(f) Storage Buildings: Buildings in which materials such as vegetables, wheat, etc. are stored is called storage building. For example, cold storage and warehouse.

(g) Hazardous Building: Hazardous and toxic substances such as uranium, acid, etc. are stored in such types of buildings.

1. Tall structures

(a) High Rise: Buildings with a height of 15m or more (regardless of their occupancy) are classified as high-rise buildings under the National Building Code of India (SP 7-2016)

(b) Skyscraper: A skyscraper is a multi-story building with at least 100m (330ft) in architectural height.

(c) Supertall Skyscrapers: Buildings taller than 300m (984 ft.) are referred to as supertall skyscrapers, whereas skyscrapers taller than 600m (1969 ft.) are referred to as mega-tall skyscrapers.

2. Bridges: A bridge is a structure designed to cross a physical barrier (such as a waterbody, valleys, roads, or railways). It is designed to facilitate passage over an obstacle that would otherwise be difficult or impossible to cross. There are several bridge designs, each having a certain purpose and being suitable under various conditions.

3. Water retaining structures: Structures that hold water either inside or outside. Water sumps, manholes, swimming pools, dams, weirs, dikes, and water tanks are just a few types of water retention structures.

4. Wind tunnels: Wind tunnels are large, air-filled tubes that are used to simulate the interaction of wind with a moving or flying object on the ground. Figure 16 shows a schematic diagram of a wind tunnel.

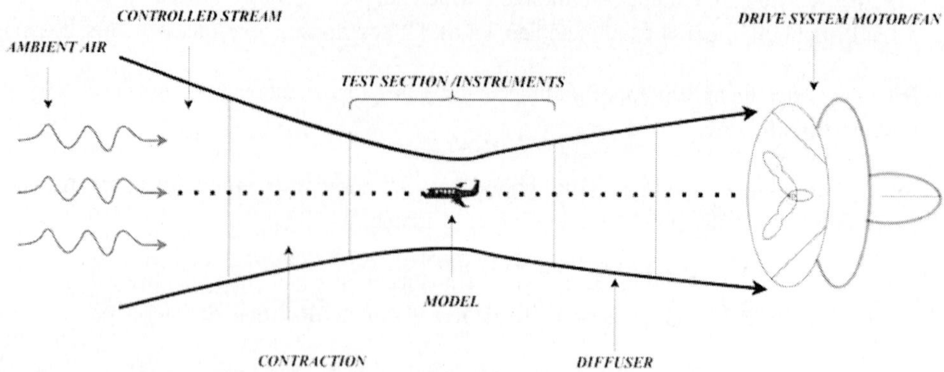

Figure 16: Wind tunnel

5. Power plant structure: A power plant or station provide the electricity that we use at home and in industries to run various machines. Nuclear power, coal combustion, water, solar, and wind are all sources of energy.

(a) Nuclear power plants: Nuclear power is the use of nuclear reactions to produce energy. Nuclear energy is generated through the processes of nuclear fission, nuclear fusion, and nuclear decay. Today, majority of nuclear power plants produce energy by plutonium and uranium fission.

(b) Hydroelectric power plants: In hydropower facilities, water flows through a pipe, known as a penstock, and then pushes against and turns turbine blades, turning a generator to generate electricity.

(c) Coal-fired power plants: Coal power plants are thermal power plants that utilize coal combustion to create energy. These thermal power stations produce one-third of the world's electricity.

(d) Geothermal power plants - Hydrothermal power plants make use of both water (hydro) and heat (thermal) resources (thermal).

(e) Solar Power Plants - The main goal of a solar power plant is to turn sunlight into electricity. This can be done directly with photovoltaics (PV) or indirectly with concentrated solar power (CSP).

(f) Wind Power plants - This is a collection of wind turbines that generate power at the same time and in the same location using the wind.

2.3. Fundamentals of Building Materials

The study of building materials is essential for all fields of civil engineering. It is important to choose the correct materials for construction based on the climate of the region, economic aspects, available resources and technology of the building industry, duration and purpose of the project, and environmental impact of the material.

India is one of the most climatically diverse countries in the world with conditions varying from dry deserts in the west, mountains, and glaciers in the north, humid tropical areas in the southwest, and the island territories. Therefore, it is important to consider the strength of the materials to withstand high and low temperatures, extreme rainfall, resistance to seawater, etc., depending on the conditions the structure is expected to endure. Availability of local materials and technology with the existing building industries plays a vital role in completing the project within a reasonable budget. The environmental cost is also a critical factor in deciding the building material for sustainable construction. The environmental impact can be beyond the production and logistics cost of the material. For example, using glass for building construction can be helpful in trapping the heat in cold weather, whereas it can drastically increase the energy requirement for cooling the building if used in hot weather climates. Building materials have also evolved over time based on the demand of the industry. Building materials are classified based on their physical and mechanical properties.

2.3.1. Physical Properties

•Density: Defined as the mass of a unit volume of homogeneous material.

•Bulk Density: It is the mass of a unit volume in its natural structure.

•Porosity: It is defined as the ratio of the volume of the pores to the volume of the specimen.

•Water Absorption: This denotes the ability to absorb and retain water. This can be expressed either by water absorption by volume (W_v) or water absorption by weight (W_w).

$$W_w = \frac{M_1 - M}{M} \times 100$$

$$W_v = \frac{M_1 - M}{V} \times 100$$

Where M_1 = mass of saturated material

M = mass of dry material

V= volume of material including the pores

•Water Permeability: It is the property of the material that allows water to penetrate under pressure.

•Fire Resistance: It is the ability of a material to resist deformation and loss of strength at a high temperature.

2.3.2. Mechanical Properties

•Strength: Strength is the ability to resist the stress applied to the material. The strength is tested in terms of compression, tension, bending, and impact. It is important to study the strength in different aspects before choosing the material. The concrete and stones are very good under compression however their strength is very low (1/5 to 1/50) under tension.

•Hardness: It is the ability of a material to resist penetration by a harder body.

•Elasticity: This defines the ability of a material to return to its original state and dimensions after the removal of the load. In the elastic state, the deformations are proportional to the applied stress.

•Plasticity: This is the ability of a material to transform its shape and size without cracking and to retain the deformed shape after removal of the load. Steel and copper are examples of plastic materials.

2.3.3. Characteristics behavior under Stress

The performance of any building material under stress can be assessed using parameters such as ductility, brittleness, stiffness, flexibility, toughness, and malleability. The ductility indicates that the material deforms significantly before complete failure under stress. Brittle materials break quickly after taking the peak load such as concrete, brick, and cast iron.

2.3.4. Type of Building Material

(a) Structural Clay Product: The clay products can be in the form of bricks, tiles, and pipes. The clay bricks can be used for structures needing aesthetics, strength, and durability. Clay tiles are used as partition walls having high resistance to fire, and clay pipes can be used in sewers and drains for their lower costs. Clay bricks are rectangular in shape having standard dimensions of 19x19x9 or 19x19x4 cm. An indent called frog (1-2 cm deep) is provided for 9 cm height brick. This frog forms a key for holding the mortar and ensuring a better connection with the other bricks.

(b) Wood: Wood is a natural polymeric material that is hard and fibrous. It is obtained from a major part of the trunk and the branches of the tree. Wood falls under two major categories a) natural such as timber, lumber, etc., and b) man-made such as plywood, fiberboards, chipboards, etc.

(c) Concrete: Concrete is made up of cement, well-graded fine and coarse aggregates, and water. Cement acts as a binding agent for all the aggregates and fills the void of fine and coarse aggregate. Water reacts with cement and increases the flow and workability of the concrete. An appropriate proportion of all the materials is used to gain the required strength. If ingredients (cement: fine aggregate: coarse aggregate) of concrete are mixed in a fixed proportion, the obtained strength of the concrete is constant and is given below:

Grade of Concrete	**M10**	**M15**	**M20**	**M25**
Mix (Cement: Fine aggregate: Coarse aggregate)	1:3:6	1:2:4	1:1.5:3	1:1:2
Characteristic Strength (MPa)	10	15	20	25

(d) Bitumen, Tar & Asphalt: Bitumen is a viscoelastic nanocrystalline by-product of petroleum. It is mainly used as an adhesive in road construction, damp proofing the basements, and painting timber and steel structural elements. Tar is obtained in the process of destructive distillation of coal, petroleum, wood, and other organic materials at high temperatures. Asphalt is a combination of an inorganic mineral matter, either calcareous or siliceous, and an organic matter.

(e) Ferrous Materials: Ferrous materials include iron, cast iron, wrought iron, and steel. Metal is extensively used in structural load-carrying members of buildings and bridges. The metal is good in both tension and compression and therefore incorporated in the form of deformed bars in the reinforced concrete. Steel is also used in trusses. Howrah bridge (Figure 17) in West Bengal is one of the country's finest and oldest steel structures.

Figure 17: Howrah Bridge Steel Structure

(f) Other building materials include ceramic material, polymeric material, paints, rubbers, glass, adhesive, etc. Ceramic materials can be used for roofs, metallurgical furnaces, refractories, etc.

2.4. Sustainability in Construction

Sustainability is a contemporary subject that has the potential to create an inspiring future for engineering, particularly civil engineering. Engineers are responsible for improving the extent to which communities can sustain soil, air, water, and energy assets, by creating a sustainable vision of the world.

The American Society of Civil Engineers defines sustainable development as "*the challenge of meeting human needs for natural resources, industrial products, energy, food, transportation, shelter, and effective waste management while conserving and protecting environmental quality and the natural resource base essential for future development.*"

Sustainability can be categorized into four types:

(a) Energy sustainability: Energy is an essential factor and engine for economic development and growth. Natural resources such as natural gas, coal, and other petroleum products are being consumed at a rapid pace and since its consumption rate is much more than their renewal rate, it is important to attenuate the usage of these energy resources to preserve their supplies for the coming generations.

(b) Ecological sustainability: To ensure the maintenance of services such as biological variety, pollination, food production, and other cleansing processes, the ecosystem's overall health must be maintained. Human activity is directly responsible for the deterioration of our planet's health. Ecological sustainability involves balancing ecological needs that are both natural and human in nature. For the foreseeable future, figuring out how ecosystems may maintain a high level of resilience while losing some of their constituent pieces will be a key topic in ecology.

(c) Soil sustainability: Soil is a nonrenewable resource that civil engineers use as one of their foundation components. For the stability of civil engineering projects such as buildings, highways, towers, tunnels, dams, etc., the foundation of soil is crucial. However, human activities such as house construction, road erosion, etc. have resulted in degrading the fertility of the soil. It is important to understand that this rapid development is a double-edged sword. Landslides are a common occurrence in hilly locations, and they are the end result of deforestation and development, in which humans invade the ecosystem and disrupt natural phenomena. All these activities, finally, disrupt the soil ecosystem and are a threat to the multi-functionality of species that are dependent on it. Therefore, it is the prime obligation of civil engineers to keep this double-edged sword of development under check.

(d) Environmental sustainability: The well-being of humans is closely linked with the health of the environment. Rapid development has led to notable changes in the environment which has long-term implications for human health. In 2019-2020, Australia saw a disastrous fire season because of rising surface temperatures induced by human activity. If more development projects do not adopt sustainability, the frequency of such forest fires is expected to increase. Another alarming environmental impact is visible frothing in the Yamuna River for several years. The requirement for achieving long-term river water quality through prioritizing biodiversity conservation is critical.

2.4.1. Evaluating Sustainable Development

Evaluation is crucial in the learning and mutual accountability processes. For sustainable development, we need to deploy technology that minimizes negative environmental impact. When it comes to addressing the issue of sustainability, the following aspects must also be considered.

(a) Resources used in the technologies: The resources used in the production must foster a shift from non-biodegradable to biodegradable materials in its production and replace nonrenewable with renewable resources. Apart from this, packaging requirements must be minimized too.

(b) Life expectancy of the project: A civil engineer must consider the project's long-term effects, whether it will last five or twenty-five years. Early material replacement or high maintenance increases resource demand unnecessarily, wasting resources that could be saved or used elsewhere. Another stage is to reduce waste, which requires consideration of the biodegradability of the materials or goods used. Furthermore, it is critical to examine the long-term impact of any project on the area's total biodiversity, as well as any potential difficulties that may arise in the future.

(c) Ecological consequences of the project: The primary purpose of civil engineers should be to ensure societal advancement as well as ensure the sustainable use of land, air, and water resources. This goal can be reached by means of managing these resources in such a way that deterioration of the environment is limited.

(d) Additional Evaluation Considerations: At first glance, determining sustainability is simple: ensuring that the natural resource in question is not exhausted. Sustainability is not viewed as a single-criterion issue in this statement. Social, political, and economic factors should be examined holistically. Other resource concerns may also play a role in the assessment. Engineers should consider a project's entire life cycle and make design changes to extend it.

2.4.2. Skills, Knowledge, and Attitudes

The discourse on sustainability is endless and encompasses a multitude of topics from all fields of human knowledge. Sustainable development education is a critical component of student education around the world because it provides the skills, information, attitudes, and values required to ensure the long-term viability of humanity.

(a) Skills: Engineers must have holistic values in order to see links between seemingly disparate components and integrate them in a way that their total worth exceeds the sum of the component values. They must also have outstanding oral and written communication abilities. The generation of fresh ideas is only the first step on the path to innovation. Engineers must comprehend the approach for promoting their ideas, attracting venture money, and strategically operating their projects while weighing the social and environmental consequences. It is critical that they seek to reduce risk factors while also managing competing project demands.

(b) Knowledge: Sustainability involves all aspects of economic, environmental, and social domains. It is imperative that engineers understand that, though their work may be regional and immediate, its likely impact will be universal and perpetual. They must exercise professional judgment and take initiative while looking at the big picture, ensuring that their solutions will improve sustainability.

(c) Attitudes: Engineers play a critical role in preventing environmental damage during construction projects. As a result, their environmental sensitivity is critical in ensuring that construction projects are completed in a sustainable manner. They must ensure that human values are considered in decision-making and that human needs are prioritized to arrive at an objective agreement on what basic human requirements are and whether they will be met for future generations while taking moral values into consideration across all subgroups.

2.4.3. Contemporary sustainable practices in civil engineering

When it comes to sustainable practices, they can be defined as those that are both environmentally benign and resource efficient. In India and around the world, the engineering sector has converted to a sustainable method in response to the need of the hour. Here are several practices that have contributed to long-term sustainability.

(a) Use of plastic in road construction: As per a government decision published in November 2015, all road developers in the country must employ waste plastic and bituminous mixes for road building. Plastic disposal has always been a source of concern. Prof Rajagopalan Vasudevan from Thiagarajar College of Engineering in Madurai is known as India's "plastic man" for inventing the revolutionary process for reusing plastic garbage to build better and more durable roadways. He received the Padma Shri Award for his contribution to sustainability. His patented invention entails combining shredded plastic with hot gravel and applying it to molten asphalt. Because plastic and bitumen are both petroleum compounds, they bond well together, increasing the toughness and the life of the roads.

(b) Switching to green concrete: The usage of green concrete (made of industrial waste and inorganic polymer) is relevant in sustainable engineering because of the amount of energy consumed in production and the harm it causes to the environment. This green concrete consumes less energy and emits less carbon dioxide than regular concrete, making it less hazardous to the environment. It is used in dams, bridges, buildings, and columns, among other things.

(c) Using harvested rainwater: Water is one of the most crucial components of the construction process, and it is also a nonrenewable resource, thus it is critical to conserve it and use it wisely. Several construction companies are already using harvested rainwater for a variety of purposes, including concrete curing, preparation, and cleaning. This is quite beneficial in conserving local water resources.

2.5. Construction and contract management

2.5.1. Construction Management

Development of infrastructure is analogous to the development of society. Infrastructural development includes the development of roads, railways, buildings, etc. The construction of these structures comes with various challenges at every phase of the construction process. The challenges might be unique according to the local conditions. The management of the construction process involves the management of large amounts of materials, equipment, manpower, time, and money. Construction management is concerned with the efficient management of the resources required in the entire construction process of a civil engineering project.

2.5.1.1. Project management

Project management deals with the management of material and manpower to increase the productivity of the construction process. Project management involves three phases

1. Project planning: Project planning is the first phase of a project which includes defining the objectives, establishing the goals and strategies for the completion of work, estimating of resources required for the project, exploring alternative strategies, and selecting the best possible combination for the timely completion of the project. Planning is however subject to change

depending on the challenges that appear during the construction work which might not be foreseeable.

2. Project scheduling: Scheduling is the process of planning the timing of the activities of the project in which they are to be performed. It involves allocating the timings of the activities, and allocation of resources as per the activity.

3. Project controlling: Controlling phase is undertaken after the project has started. Controlling is the process of determining deviations from the original plan and re-plan to compensate for the deviations.

Project management essentially requires the project manager to efficiently complete the project with the help of proper tools like bar charts and milestone charts, network diagrams, etc.

2.5.1.2. Quality management:

Maintaining the quality of construction is the most important part of construction management so that the structure is able to provide service during the period it is designed for. A major part of the work in construction is done on the ground and hence maintaining the quality of the work is dependent on the performance of individuals involved in the construction. Hence everyone must be made aware of the quality standards expected in the construction of the project. To maintain proper quality standards, the following steps are followed:

- Define the standards of design and construction of the structure.

- Define the procedure and standards of tests and monitoring to be followed during the construction work.

- Define the role of administrative and management departments in ensuring the proper quality standards.

Routine inspection and supervision of the predefined procedure and standards are required for maintaining the quality prescribed in the design. The inspection involves making sure that the set parameters are satisfied with regards to geometry, strength, appearance, etc.

Quality must be maintained right from the start to the end of the project. Any compromise in any step of the process will lead to the possibility of failure of the structure.

2.5.1.3. Safety management:

Construction workers are exposed to various safety hazards, for example, a worker working on a high-rise structure is at risk of falling or some construction equipment like nuts, bolts, wrench, etc. might fall on a worker standing below it. While working on a structure built on water bodies, like bridges, workers have the risk of falling into the water. There are accidents that damage the structures like a failure due to overloading during construction, failure of scaffoldings, etc. These accidents result in the loss of lives along with monetary losses. These accidents and occupational hazards are taken into serious consideration in civil engineering and hence are discussed in the course.

The standard practices of safety against the risk of accidents are to be employed and must be enforced by the supervisor in charge of the construction site, much like that in the case of quality control. The role of the supervisor is crucial for maintaining proper safety standards in the construction site. The people must be properly trained such that they should know the use of proper equipment and the possible hazards and risk of the work undertaken by them. There must be the proper facility of safety equipment such as safety nets, warning boards, shielding the sharp edges, etc., and personal safety gear for the workers like helmets, steel toe boots, etc. Routine supervision of safety measures must be undertaken to detect unsafe conditions and correct them for the smooth functioning of the project and the safety of the workers and the structure.

2.5.2. Contract Management

Every project requires certain quality control measures, compliance with rules and regulations along with supervision to achieve the desired result. In this, contract management plays an important role in managing all the stakeholders such as vendors, partners, customers, or employees. A contract is a document explaining the agreement between the two parties regarding resources, relationship, and risk. It is an agreement that defines the roles and responsibilities of different stakeholders involved in the project and the burden of mitigating the risk.

2.5.2.1. Importance of Contract Management

A construction contract agreement is a document defining the terms of a construction project. The parties involved who must agree to the terms written are the General Contractor and the client/owner. Any contract agreement contains:

1. Duration of the project or construction schedule reviewed through progress meetings and updated from time to time

2. A statement of work defining who agrees to do what, the technical specification for carrying out the work, and material compositions.

3. Describes the rights and obligations of shareholders and sets forth the procedures to be followed for settlement in the event of any disagreement.

Conception, design, pre-construction, construction procurement building, and delivery/post-construction are the steps/phases that make up a construction project. All of them fall under the umbrella of contracts, which can be broadly classified into pre- and post-award contract development. As soon as the bid is accepted, construction starts. As the project advances contract changes are likely to take place and need to be efficiently handled and incorporated into the main contract. Ideally, the original documentation would have already described how this procedure works.

2.5.2.2. Important Elements of Contract Management

It is not sufficient for a company to have people in place to manage construction contracts. Employees must be supported via means of procedures and software partners to fulfill the expanding compliance and analytical requirements. When a contract management plan is efficiently put into practice, organizations can anticipate seeing the following:

1. The anticipated business advantages and commercial returns are being fulfilled.
2. The supplier is helpful and receptive to the organization's requirements.
3. The organization experiences no contract conflicts or unpleasant surprises.
4. The service delivery meets everyone's (parties involved) expectations.

2.5.2.3. Stages of Contract Management Process

To organize work and manage the typical contract process, contracts are divided into stages. The process in the stages of contract management includes basically the creation, collaboration, signing, tracking, and renewal. The stages are explained in the following steps:

(a) Initial requests: Finding contracts and pertinent documents that support the goals of the contract is the preliminary advancement.

(b) Authoring contracts: Handwritten contracts take sufficient time to surface, but the procedure may be made much more effective with the use of automated contract management tools.

(c) Negotiating the contract: Once the contract is drafted, personnel may be able to compare copies and highlight any differences to reduce the need for further discussion.

(d) Approving the contract: The bulk of obstacles happen on getting management's permission. Users can avoid this by creating unique approval workflows that involve serial and parallel approvals to maintain speedy decision-making.

(e) Execution of the contract: Users can control and expedite the signature process when completing the contract by using electronic signatures and emails/fax capabilities.

(f) Obligation management: To ensure that important stakeholders receive deliverables on time and the contract's value does not decrease during its early development, extensive project management is required.

(g) Revisions and amendments: It's a tough task to acquire all of the documentation needed for the preparation of the first draft of the contract. When issues are uncovered that were previously disregarded, protocols must be in place to amend the original contract.

(h) Auditing and Reporting: Contract management entails more than merely drafting a contract and then filing it away without further thought. Contract audits are essential for confirming both companies' adherence to the terms of the agreement as well as any potential problems.

(i) Renewal: Manual contract administration practices regularly cause organizations to miss out on renewal opportunities and lose money. By automating the procedure, a company can locate chances for renewal and create new contracts.

2.6. Repair and rehabilitation of structures

Damage and decay of structures are unavoidable. This occurs due to the deterioration of materials, repeated loading, and exceptional events. However, their effects can be tolerated as long as the structure meets requirements related to structural capacity, aesthetics, comfort, and safety of the user, and economic and market values (Lourenço et al., 2022). When such requirements are no longer met, there arises the need to perform repair or rehabilitation of the structure.

Such repair and rehabilitation of existing infrastructure have become increasingly important due to 1) the construction industry having been identified as a core contributor to environmental pollution as well as climate change, and 2) existing building stocks are becoming old and require measures to prolong their service life or to make them more resistant to extreme hazards such as earthquakes, fire, and blast.

Repair and rehabilitation measures are most effective when structural damage and its underlying cause are reliably identified. Similarities between the diagnostic processes of humans and building pathologies have led to a medical analogy for structural repair and rehabilitation. Such similarities start from the fact that inspection and diagnosis techniques play a major role in both, providing information and allowing the identification of adequate remedial measures. The diagnosis process can be performed in two ways Destructive Testing and Non-Destructive Testing (NDT). The NDT is preferred in repair and rehabilitation measures as the specimen does not suffer any physical damage in the testing procedure. The NDT tests such as Rebound Hammer Test, Penetration and Pull-out techniques are used to assess the strength of the concrete. The radioactive and nuclear NDT methods use X-ray and Gamma rays to measure the density and thickness of the concrete. The magnetic NDT tests are used to identify the position of the reinforcement in the concrete and other electrical NDT tests such as microwave absorption techniques are also used to measure the moisture content in the structure.

If these are performed incorrectly, the adopted retrofit measures will be ad-hoc. The probability of failing to meet performance requirements (sickness in the case of humans) can also be reduced by planning routine maintenance or monitoring (health checkups in the case of humans). Thus, the recommended methodology for repairing/retrofitting structures is as highlighted in Figure 18. The sequence of activities to be carried out can be summarized as

1. The current performance requirements of the existing structure should be carefully

evaluated.

2. After the evaluation of the required performance requirements, a thorough inspection of the structure should be carried out.

3. The structural performance should be then evaluated. This evaluation should incorporate all findings of the inspection carried out in step 2. The evaluated performance should be then compared against the requirements evaluated in step 1.

4. If the structure satisfies its performance requirements, then its use can be continued.

5. If the structure does not satisfy its performance requirements and its continued use is desired, the engineer should proceed with designing the retrofit.

6. While designing the retrofit, an appropriate retrofitting method should be first selected. The design and implementation should be done as per technical norms and manufacturer specifications addressing it.

7. The performance of the retrofitted structure should be evaluated again to verify that the retrofitted structure meets the performance requirements.

8. Only once it is confirmed that the existing structure can satisfy the performance requirements

by adopting the selected retrofitting method, should the retrofit be implemented.

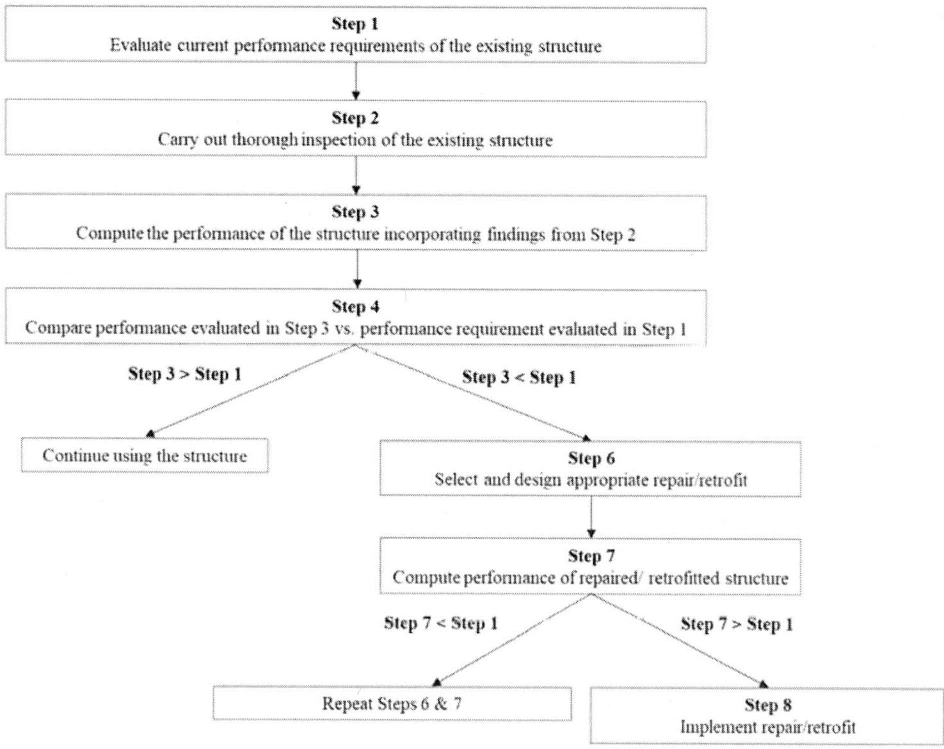

Figure 18. Conceptually, repair/retrofit methods can again be broadly classified as (Macchi et al., 2018)

•Adding new structural elements: New structural elements could be added to help the existing building better sustain loading actions. For example, newly inserted elements in buildings could be steel braced frames or new walls while for bridges, this could mean the insertion of new brick piers.

•Strengthening existing elements: Retrofit/repair strategies could also be based on the strengthening of the existing structural elements. For example, columns, beams, and even bridge piers can be wrapped in external jacketing composed of carbon, glass fibers, or even steel plates.

•Locally increasing deformation capacity: Such strategies address only the most critical zones of the structure identified when evaluating its performance. Examples of such strategies again include fiber wrapping or steel encasement of columns.

•Reduce the action/ demand on the structure: Retrofit/repair strategies can also operate by reducing the action/demand on the structure rather than increasing its capacity. For example, in the case of bridges, traffic restrictions could be imposed to reduce the load the bridge has to carry.

Practically, a plethora of local and global methods and products are currently available. As an example, for reinforced concrete structures not meeting performance standards under earthquakes, local strategies would include measures addressing column strengthening, beam strengthening, or beam-to-column joint strengthening while global strategies would include measures such as the addition of new walls/bracing systems, reduction of mass/ structural irregularities or the use of energy dissipation/isolation devices (Christopoulos et al., 2006). It is to be noted here that the choice of the most appropriate retrofit method depends entirely on the problem at hand. This is also highlighted in the case study addressing the dome of the presidential palace of India i.e., Rashtrapati Bhawan (Aranha et al., 2019). Such special structures rarely fall under the direct purview of any guideline and the study highlights the importance of detailed investigations that need to be carried out to evaluate the capacity of the structure in its existing state. During a visual inspection of the double dome structure of Rashtrapati Bhawan, meridional cracks were spotted in two out of the four niches in the drum of the lower dome. The complex layout of the structure, variety of materials used, and no prior knowledge of reinforcements prompted the use of multiple approaches for the detailed study and investigations of the structure. These detailed investigations are then used to propose a non-invasive as well as a reversible retrofit strategy for the dome, qualities which are both necessary considering the historical importance of the structure.

Structural engineering is a vital part of civil engineering that deals with the design and construction of civil engineering structures. Structural engineering employs basic concepts of mechanics for analyzing structures. Hence a strong fundamental understanding of the mechanics of structures is necessary for the appropriate design of structures. Infrastructural development is directly linked with the development of a society. Structural engineering is dynamic in nature and has pushed the limits of infrastructural development and the changing climate the current need for infrastructure development is to be sustainable.

Table 1: List of the equipment used at the construction site of civil engineering.

Category/ Field	Name(s)	Purpose
Earth Moving	Excavator	The primary purpose of excavator is to excavate however, it can be used for lifting heavy material and demolition purposes.
Earth Moving	Backhoe	This is majorly used for digging holes and trenches. This can also be used for excavation purpose up to 3-meter depth.
Earth Moving	Bulldozer	These have easy maneuverability in the difficult terrain of the construction site. This is used to remove and push the debris or soil on the site. This can also be equipped with a ripper which can be further used to break the hard surfaces of concrete or asphalt.
Hoisting Equipment	Tower Crane	Tower cranes are used to erect the big structures and materials such as pre-stressed concrete slab, steel frames to the required height.
Hoisting Equipment	Telehandlers	They are used to lift heavy materials and provide the platform for the workers at greater heights.
Hauling Equipment	Dumper	To transport the material in bulk. It has hydraulic system which can lift and dump the material quickly.
Hauling Equipment	Trailer	The trailers are used to shift the materials like reinforcement rods at construction sites.
Hauling Equipment	Concrete Mixer Truck	To transport the ready-mix concrete from plant to the construction site.
Road Construction	Paver	This is used in asphalt pavement laying.
Road Construction	Compactors	Compactors are used to compact the soil and earth materials.
Foundation	Pile Boring Equipment	Pile boring equipment are used to dig deeper in the soil and place the precast piles or construct in situ pile foundation.

2.7. Recycling

Construction needs large number of natural resources and generates millions of tons of waste in the form of concrete, asphalt, steel, tyres, bricks, fly ash, slag etc. Most of the generated waste is deposited in the landfills, water or air. Reutilization of these materials is utmost important for sustainable, environment friendly and economic development. The recent development has focused on the usage of recycled materials and analysis of their compatibility and suitability of their strength in building materials. The recycled aggregates can be formed from the water quenched water furnace slag and reclaimed asphalt pavement. The pervious concrete is made using the recycled brick aggregates. The agricultural by products and used tyre rubber can also be used as cement mortars.

References and suggested readings

1. "Roads Made of Plastic Waste in India? Yes! Meet the Professor Who Pioneered the Technique." The Better India, 2 Feb. 2016, www.thebetterindia.com/43685/plastic-waste-in-road-construction-plastic-man-india-prof-vasudevan/.
2. Aranha CA, Menon A, Sengupta AK. (2018). Determination of the causative mechanism of structural distress in the presidential palace of India. Eng Fail Anal 2019; 95:312–31. https://doi.org/10.1016/j.engfailanal.2018.09.023.
3. Center Of Gravity. (n.d.). https://sites.google.com/site/rpfypaerodynamics/center-of-gravity
4. Ceschinski, H & Henkes, Hans & Weinert, H & Weber, Werner & Kühne, D & Monstadt, Hermann. (2000). Torquability of microcather guidewires: The resulting torsional moment. Bio-medical materials and engineering. 10. 31-42.
5. Christopoulos C, Filiatrault A, Bertero V V. (2006.). Principles of passive supplemental damping and seismic isolation. 6th ed. Pavia, Italy: IUSS Press.
6. Danilina, Elena Ivanovna, and Vladimir Evgenevich Chebotarev. (2017). "Comprehensive Assessment of Road and Communal Infrastructure as an Important Tool for Sustainable Development of the Urban Economy." JSTOR, www.jstor.org/stable/10.2307/26234013?refreqid=search-gateway.
7. Lourenço PB, Barontini A, Oliveira DV, Ortega J. (2022). Rethinking preventive conservation: Recent examples. Geotech. Eng. Preserv. Monum. Hist. Sites III, London: CRC Press; p. 70–86. https://doi.org/10.1201/9781003308867-4.
8. Macchi G, Calvi GM, Sullivan TJ. Structural Strengthening and Retrofit; Motivations, Concepts and Approaches. (2018). p. 1–24. https://doi.org/10.1007/978-981-10-5858-5_1.
9. McCuen, R. H., Ezzell, E. Z., Wong, M. K. (2011). Fundamentals of Civil Engineering: An Introduction to the ASCE Body of Knowledge. United States: CRC Press.
10. Seismic Assessment and Retrofit of Existing RC Buildings: Case Studies from Degenkolb Engineers. SP-297 Seism. Assess. Exist. Reinf. Concr. Build., American Concrete Institute; 2014. https://doi.org/10.14359/51686906.
11. Seismic Evaluation and Retrofit of Existing Buildings. Reston, VA: American Society of Civil Engineers; 2014. https://doi.org/10.1061/9780784412855.

12. Seismic Rehabilitation of Existing Buildings. Reston, VA: American Society of Civil Engineers; 2007. https://doi.org/10.1061/9780784408841.
13. The National Assembly for Wales. (n.d.). https://www.dqi.org.uk/case-studies/heritage-culture/thenationalassemblyforwales.php
14. Thompson, P. (2010). "Teaching Sustainability in Civil Engineering Using Ceequal." Proceedings of the Institution of Civil Engineers - Engineering Sustainability, vol. 163, no. 4, pp. 209–217., doi:10.1680/ensu.2010.163.4.209.

Dynamic QR Code for Further Reading

Understanding Shear Force and Bending Moment Diagrams	BENDING MOMENT
Videos about various construction and other civil engineering work: https://practical.engineering/	PRACTICAL ENGG
How are skyscrapers built?	SKYSCRAPERS

Indian Standard (IS) codes for cement and concrete	IS CODES
Indian Standard (IS) codes for structural engineering and structural section	IS CODES
National building code for India	BUILDING CODE

3 Geotechnical Engineering

Unit specifics

Through this unit we have discussed the following aspects

- *Introduction to soil mechanics*
- *Basics of various index properties and engineering properties of soil*
- *Basics rock mechanics and engineering geology*
- *Introduction to foundation engineering*
- *Introduction to underground structures*
- *Brief discussion on solid waste management*

The practical applications of the topics are discussed for generating further curiosity and creativity as well as improving problem solving capacity.

Besides giving a large number of multiple-choice questions as well as questions of short and long answer types marked in two categories following lower and higher order of Bloom's taxonomy, assignments through a number of numerical problems, a list of references and suggested readings are given in the unit so that one can go through them for practice. It is important to note that for getting more information on various topics of interest some QR codes have been provided in different sections which can be scanned for relevant supportive knowledge.

After the related practical, based on the content, there is a "Know More" section. This section has been carefully designed so that the supplementary information provided in this part becomes beneficial for the users of the book. This section mainly highlights the initial activity, examples of some interesting facts, analogy, history of the development of the subject focusing the salient observations and finding, timelines starting from the development of the concerned topics up to the recent time, applications of the subject matter for our day-to-day real life or/and industrial applications on variety of aspects, case study related to environmental, sustainability, social and ethical issues whichever applicable, and finally inquisitiveness and curiosity topics of the unit.

Rationale

This fundamental unit on geotechnical engineering will help students to get a primary idea about the fundamental principles and theories of soil mechanics, rock mechanics, foundations, and solid waste management. It explains the various properties and characteristics of soil that are relevant in civil engineering. A discussion on rock mechanics and engineering geology is also put forward. This chapter then explains the various types of foundations and the conditions for selecting a suitable foundation for a structure. The basics of underground structures is also discussed in this chapter. This chapter lastly describes the importance and various techniques of solid waste management. The topics are discussed to a length so as to develop the basic idea of the subject. Various practical examples are mentioned in the chapter for creating a clear picture of the application of the subject in the field of civil engineering.

Geotechnical engineering is a branch of civil engineering which deals with the soil and rock surface of the earth and its ability to withstand load coming from the superstructure. Before design and construction of any civil engineering work, assessment of soil properties and its bearing capacity must be carried out and these are covered in geotechnical engineering

Pre-requisites

Mathematics: Coordinate Systems (Class XII)

Physics: Mechanics (Class XII)

Unit outcomes

List of outcomes of this unit is as follows:

U1-O1: Understand basic properties and characteristics of soil.

U1-O2: Understand the various rock structures and geological formations relevant to civil engineering.

U1-O3: Understand the basic concepts and principles of foundations.

U1-O4: Understand the basics of different underground structures.

U1-O5: Understand the various techniques of solid waste management.

Unit-Outcomes	EXPECTED MAPPING WITH COURSE OUTCOMES (1- Weak Correlation; 2- Medium correlation; 3- Strong Correlation)					
	CO-1	CO-2	CO-3	CO-4	CO-5	CO-6
U1-O1	3	3	3	-	3	1
U1-O2	1	1	2	2	1	-
U1-O3	2	1	3	1	2	1
U1-O4	-	-	3	1	2	2
U1-O5	3	3	3	-	3	1

It takes thousands of years for a boulder to break down due to weathering. The boulders are broken down into sands, silts, and clays and the composite of these particles makes up the soil. Soil is a complex conglomeration of inorganic matter, organic matter, water, and air in varying proportions. All civil engineering structures eventually transfer the load to soils, either directly or through masonry. Knowing the expected response due to the imposed loadings is of primary interest to civil engineers. Geotechnical engineering is the specialist branch that deals with the study of the properties of soils and the impact of loading forces and soil-water interactions on the behavior of soils.

3.1. Soil Mechanics

Soil mechanics is the application of the laws of mechanics and hydraulics to sediments and other deposits. The broad area of "soil-structure interactions"in geotechnical engineering represents the interactions between the three linked systems of any civil engineering structure that is built on or inside the ground: the structure, the foundation, and the soil underlying and surrounding the foundation. Its common applications include the design of pavements, retaining walls, dams, embankments, foundations, and underground structures such as tunnels, water distribution pipes, etc. Geotechnical studies are crucial for mitigating risks and the safe design of any civil engineering structure.

3.1.1. Determination of texture of soil

Soil texture (such as loam, sandy loam, or clay) is the characteristic that indicates the relative content of particles of large (sand), medium (silt), and small (clay) sizes. Because of its smaller size, clay clumps together easily and is cohesive in nature while sand is non-cohesive in nature.

Sand is visible to the naked eye and is pervious. Clay particles are micron-sized particles less than 2 micron. Loamy soil is a mixture of clay, silt, and sand in proportions of 20%, 40%, and 40 % respectively. Various types of soils have different benefits, for example, crops require soils with high organic matter. Figure 19 shows the types of soil particles along with the arrangement of the particles and the particle size. By seeping downward, water breaks down rocks into the soil and dissolves them into the ground.

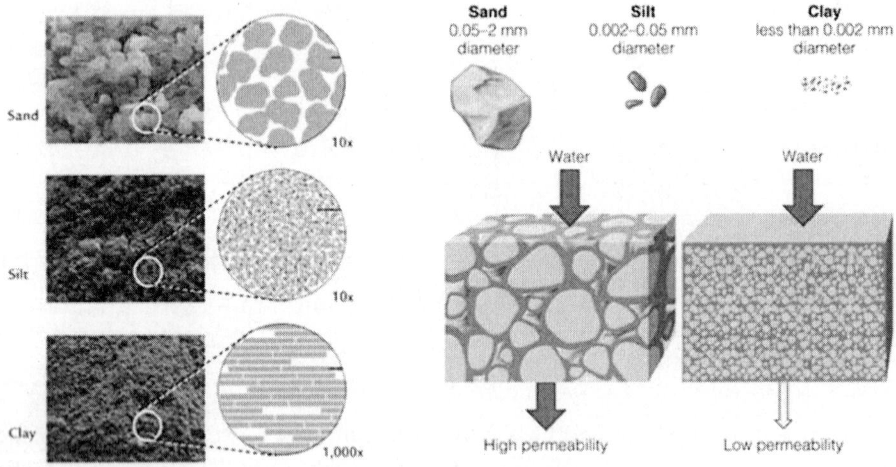

Figure 19: Types of soil particles (Source: Thomas higher education)

The following field tests are used to determine the texture of the soil. These simple tests require fine earth, which is a mixture of sand, silt, and clay. For the test to be accurate, the fine soil particles should first be separated from larger particles such as gravel and stones.

3.1.1.1. Throw ball test

The soil can be tested by squeezing some moist soil into a ball and throwing it about 50 cm in the air and catching it. If the ball falls apart, it indicates that the soil has more sand and if it sticks together, the soil has more clay, as shown in Figure 20.

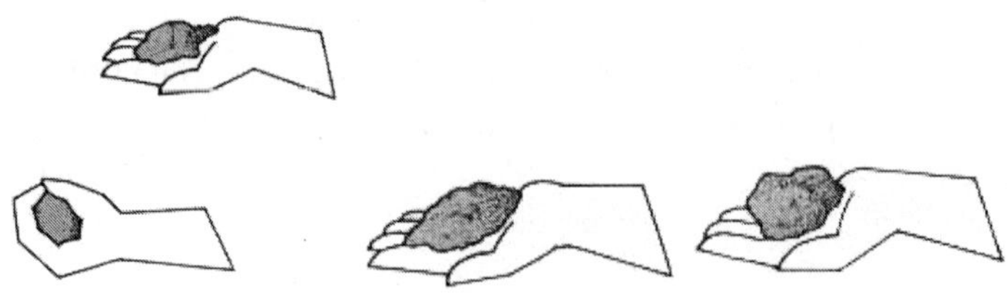

Figure 20: Throw ball test (Source: Food and Agriculture Organization, United Nations)

3.1.1.2. The bottle test

In this test, 5 cm of soil is put into a bottle and filled with water. The soil and water are then stirred together, and the bottle is left undisturbed for an hour. After an hour, the water gets clear and larger sized particles settle. Organic matter floating on the surface of water indicates a clay layer. After an hour, if the water has still not cleared up, it indicates the presence of fine clay. If the water is still mixed with soil that forms a layer in the middle, it indicates silt while the bottommost layer is sand. The proportion of each can be determined by measuring the depth of sand, silt, and clay. This will provide an approximate idea of the proportion of sand, silt, and clay in the soil (Figure 21).

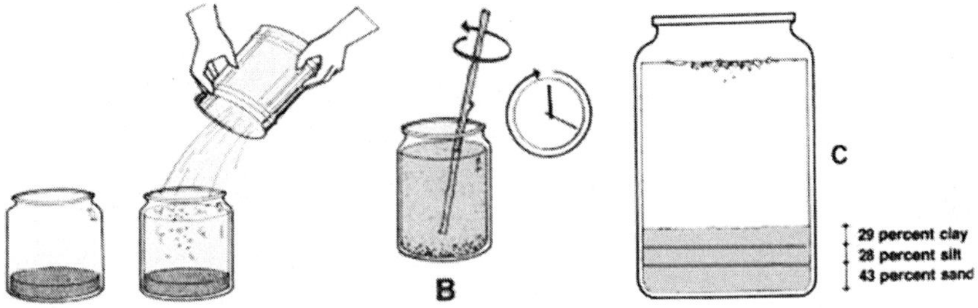

Figure 21: Bottle test which determines proportion of sand, silt, and clay (Source: Food and Agriculture Organization, United Nations)

3.1.2. Formation of soil

Organic and mineral matter in soils can be found in solid, gaseous, or aqueous states. Soil is formed due to chemical weathering processes such as oxidation, reduction, carbonation etc. as well as physical weathering of rock particles due to changes in temperature, abrasion etc. The physical characteristics and the chemical properties of soils include cohesion, adhesion, color, texture, specific gravity, and mineral content. Residual soil accumulates or deposits near the parent rock. whereas transported soil is deposited away from the origin of formation. There are five types of transported soil, according to the medium of transport:

(a) Alluvial

(b) Aeolian

(c) Glacial

(d) Lacustrine

(e) Marine.

Alluvial soils are formed by accumulated sediments transported by flowing water. Aeolian soils are formed from sandy parent material transported by wind. Glacial soils are deposits transported by ice and found in regions having rocky terrain with ice blocks. Lacustrine soils are depositions in the bottom of ancient lakes, typically formed by sediment carried by river or stream channel to the lakes. Finally, marine soil are sediment deposits on the sea or ocean floor whose texture and composition depends on proximity to land and organic matter.

3.1.3. Soil deposits in India

Based on origin, color, composition, and location, the soil deposits in India can be broadly classified into five types. Figure 22 shows the map of major soil types found in India.

(a) Alluvial soil

Predominantly found in northern and eastern India, typically in the river valleys of the Ganga and the Brahmaputra, the distinctive feature of these soils is that they are composed of layers of alternating sand, silt, and clay. Generally, the size of the layer depends on the floodplain characteristics. Alluvial soils have one of the highest agricultural productivities among all soils, requiring the least water due to its high porosity.

(b) Black cotton soil

These soils are found in the central and southern region (Deccan plateau) of India. These are residual soils of basalt rock mostly used for cotton cultivation. These soils are highly susceptible to swelling and shrinkage and have low bearing capacity due to montmorillonite clay mineral.

(c) Laterite soil

This is a residual soil formed from basalt that is characterized by reddish or pink colors due to the presence of iron oxides and aluminum oxides. These soils are found throughout central, southern, and eastern India.

(d) Desert soils

These soils are transported by wind and are highly pervious and non-cohesive. The size of these soil particles is coarse grained and requires compaction to increase strength and bearing capacity.

(e) Marine soil

A narrow belt of marine deposits mainly lies along the coast of India, and these deposits consist of sand deposited above deep deposits of soft marine clays. These soils are highly compressible, soft, and plastic in nature due to the presence of organic matter and have low strength.

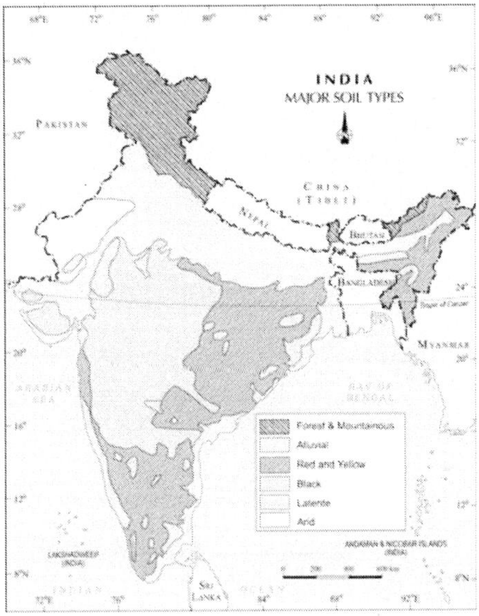

Figure 22: Major soil types of India. (Source: Wikipedia)

3.1.4. Classification of soil

Soil is classified based on certain distinguishable properties. In geotechnical engineering, the classification of soil is based on the engineering properties that are related to civil engineering applications. There are different classification systems put forward by various organizations. Some of the classification systems that are relevant for geotechnical engineering are discussed below.

1. Textural classification of soil:

Textural classification system is one of the most widely used classification systems developed by the United States Department of Agriculture (USDA). The classification is based on the percentage of sand, silt, and clay in the soil sample. The percentages are plotted on the three sides of a triangle as shown in Figure 23 and the type of soil can be determined as per the percentage of sand, silt, and clay present.

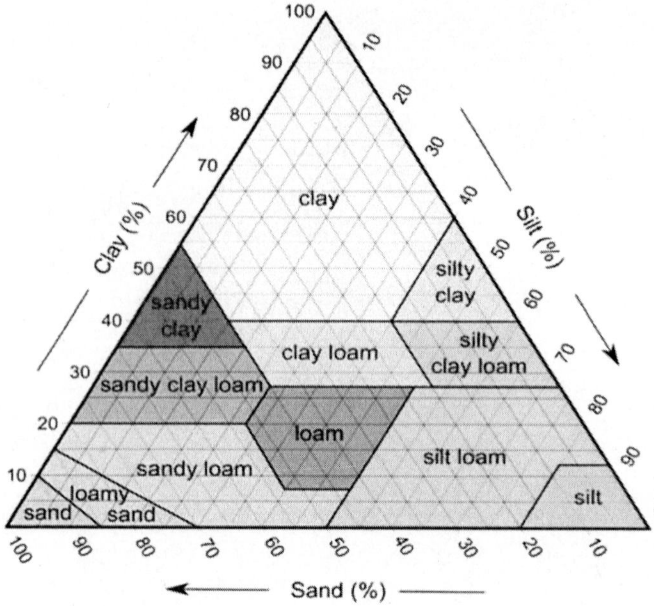

Figure 23: A ternary diagram of the soil texture triangle (Source: United States Department of Agriculture - USDA)

2. Unified soil classification system (USCS):

USCS is a widely used and universally accepted system of classification for application in geotechnical engineering. The classification is based on the soil particle size and the plasticity characteristic of soil. The soil is broadly classified as coarse grained, fine grained, and organic soil.

(a) Coarse grain soil: According to USCS, soil is classified as coarse grained if 50% or more soil is retained on a sieve size of 0.075mm. Coarse grained soil is further subdivided into gravel (G) or sand (S) depending on whether 50% or more coarse grained soil is retained or passed through on a sieve of 4.75mm, respectively.

(b) Fine grained soil: If 50% or more soil passes through a sieve size of 0.075mm then it is termed as fine-grained soil. Fine grained soil is classified as silt if the plasticity index lies below the A-line, otherwise it is silt.

The A-line is shown in Figure 24. If the liquid limit is more than 50% then it is termed as highly plastic, otherwise it is termed as low plastic soil.

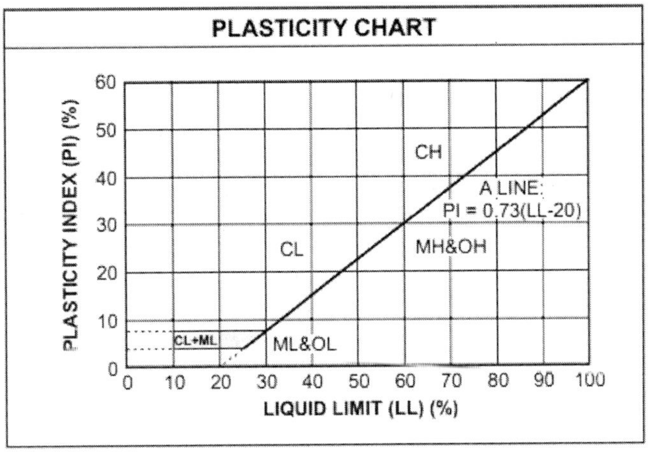

Figure 24: Plasticity Cart for United Soil Classification System (USCS) (Source: https://dot.ca.gov/).

(c) Organic soil: Organic soils are fine grained soils that contain substantial amounts of organic particles and exhibit organic properties such as strong odor, dark color, etc.

3. Indian standard system of soil classification (ISSCS):

ISSCS adopts a method similar to USCS for classification of soil. The difference is in the distinction of plasticity. Unlike USCS, soil is considered low plastic if liquid limit is below 35%, medium plastic if liquid limit is between 35% to 50% and if the liquid limit is above 50% it is considered as highly plastic.

3.1.5. Phase diagram

A phase diagram is a diagrammatic representation of soil which consists of solids, water, and air. It is classified into a three-phase or two-phase system. We consider fully saturated soil as a two-phase system where air voids are completely replaced by water, i.e. degree of saturation is 100%. While partially saturated soil is a three-phase system consisting of solid, water, and air. Lastly dry soil is a two-phase system in which voids filled with water are completely replaced by air or we can say that no water is present in voids.

3.1.6. Volumetric relationships

To analyze the behavior, volumetric, relation and properties of soil we study 3 phase diagrams of soil (Figure 25). Mostly 5 volumetric relationships were used in soil mechanics for engineering purposes such as void ratio, porosity, air content, degree of saturation, percentage of air void.

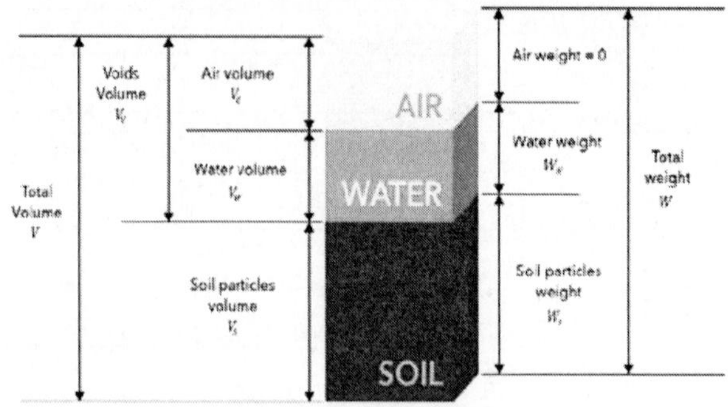

Figure 25: Three Phase diagram (Source: https://edtech.engineering.utoronto.ca/)

(a) Void ratio.

It is the ratio of volume of voids to volume of solid soil. Void ratio has higher value for fine grained compared to coarse grained and its value can be greater than one.

(b) Porosity.

It is the ratio of volume of voids to total volume of soil mass. Its value can't exceed 100%. Void ratio and porosity measure the degree of compactness. Dense soil has less void ratio and porosity.

(c) Degree of saturation.

It is the ratio of volume of water to volume of void. A soil is considered as dry soil when degree of saturation is zero and fully saturated when degree of saturation is 100%.

(d) Percentage air void

It is the ratio of the volume of air to total volume.

(e) Air content

It is the ratio of volume of air to volume of void. If the soil is fully saturated, air content and percentage air void is zero.

3.1.7. Index properties of soil

The term "soil index" refers to properties of soils that facilitate identifying and classifying them for engineering purposes. Plastic soils are cohesive in nature such as clay and silt and granular soils such as sand and gravel are cohesionless. Plasticity of cohesive soil is given by the plasticity index. Index properties for fine grained soils are consistency, clay mineralogy, structure, shape, water content. Similarly, index properties for coarse grained soil are relative density and clay minerals content.

(a) Clay mineral and structure

Clay minerals are very small crystalline substances which are the basic components of clay material. Clay minerals are very small in size and flaky in shape and are formed because of the chemical weathering of rocks. They are visible only under an electron microscope. The clay minerals can be divided into three main groups based on the arrangement of the crystalline structure, namely, kaolin, montmorillonite, and illite. Minerals within a group exhibit similar properties. The atomic structure of clay minerals basically consists of two types of crystal sheets - the tetrahedral or silica sheet and octahedral or alumina sheet. The arrangement of the sheet, the nature of bonding and the metal ions present in the crystal lattice makes the different clay minerals.

(b) Consistency

Consistency depends upon water content and describes the degree of firmness in a soil and indicates the soil to be soft or stiff. The water content at which soil changes from one phase to another is termed as a consistency limit. Figure 26 shows the different consistency limits for different water content. Soils with high water-content have zero shearing resistance. Therefore, soil at liquid phase exhibits negligible shear strength. In the plastic phase, the soil can be easily molded, despite some shear strength, but at the same time it is easy to fracture in the semi-solid phase, as the moisture content is reduced. This index property holds true for cohesionless soil as well as cohesive soil. Cohesionless soil mainly depends on grain size and shape of particles and cohesive (plastic) soil mainly depends on water content.

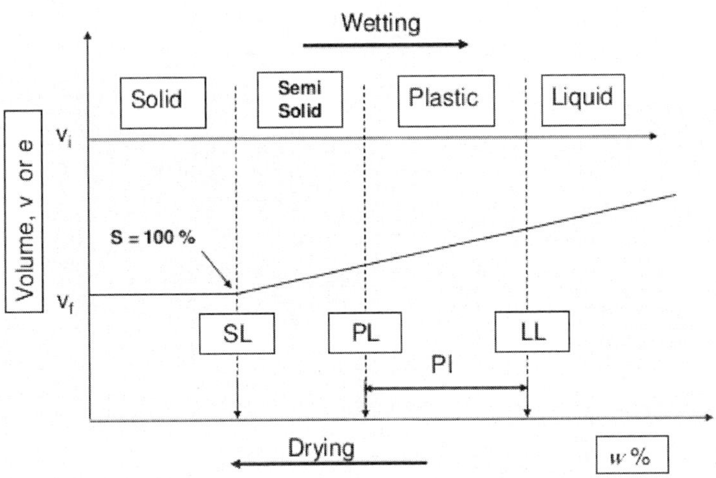

Figure 26: Atterberg's limit (Source: Civil Engineering, Texas Tech University)

•Plastic limit (PL)

Moisture content at which soil just starts behaving as a plastic material above the plastic limit soil is in plastic state and below plastic limit soil is in a semisolid state and as it reaches semi solid state, the soil loses its plasticity.

•Liquid limit (PL)

Liquid limit is the point at which the behavior of soil changes from a liquid state to a plastic one based on the amount of water it absorbs. When this happens, the soil exhibits a low shear strength of 2.7 N/mm^2. The liquid limit of soil is determined by the clay mineral. The stronger the surface charge and the thinner the particle, the greater the amount of water it will soak up. It is determined by using the Casagrande apparatus.

•Shrinkage limit (SL)

Shrinkage limit is the water content at which soil stops behaving like a solid material. Reduction of water content below shrinkage limit signifies no change in volume.

Liquidity index

Liquidity index of soil is a measure of its water content in relation to its liquid limit. When soil is at its liquid limit, its liquidity index is 100% and the soil behaves as a liquid. At the plastic limit, the soil has a zero-liquidity index. Negative values indicate a water content less than its liquid limit.

(c) Plasticity index

It is defined as the difference between LL and PL, the range at which soil remains in a plastic state. Plasticity index is zero when plastic limit (PL) exceeds the liquid limit (LL)

(d) Consistency index

It indicates the amount of moisture in a soil in relation to the plastic limit. For example, zero consistency index describes a very soft and low-shear soil. A soil with a consistency index equal to the plastic limit indicates that the soil is reasonably firm. It is possible to obtain a negative value for consistency index, which reflects higher water content percentage than liquid limit. As the soil is semi-solid, this indicates that the percentage of water content is comparatively more than liquid limit.

(e) Flow index

The slope of the flow curve obtained by Casagrande between water content and number of blows, mostly water content corresponding to 25 number of blows, is considered for analysis. With increase in water soil losses its shear strength, according to flow curve steeper slope indicate lower shear strength compared to flatter slope.

(f) Toughness index

It is the ratio of plasticity index to flow index and measures the shear strength at the plastic limit. Its value lies between 0 to 3. If a soil has a toughness index less than 1 then soil is friable.

(g) Activity

It is the ratio of plasticity index to the percentage of clay particles finer than 2 microns. Water holding capacity of soil increases due to clay minerals. With higher value activity, soil is expected to show characteristics of a clayey soil like high swelling or shrinkage, high compressibility etc.

(h) Sensitivity

It is the ratio of unconfined compressive strength of undisturbed soil to unconfined compressive strength of remolded soil with the same moisture content. It measures the loss of shear strength due to remolding or reorientation. Its value lies between 1 to 16.

(i) Thixotropy

It is the property of soil to regain the shear strength of remolded soil with time.

3.1.8. Engineering properties

The main geotechnical properties of soil are permeability, compressibility, and shear resistance. Settlement of buildings built on soils is calculated by considering the compression characteristics. Soil strength determines slope stability, soil bearing capacity, and earth pressure on retaining structures.

(a) Shear strength

Shear strength of soil is the ability to withstand tangential stress. Shear failure of soil mass occurs when the applied compressive loads generate shear stresses that exceed shear strength. Shear strength influences the stability of the soil mass under loads by controlling the capacity of soils, slope stability, and the lateral earth pressure on walls and retaining walls

(b) Compressibility and compaction

Compressibility indicates the capability of soil to decrease its volume under loads. Expulsion of water through voids due to static action is termed as consolidation while expulsion of air through voids due to dynamic action known as compaction. Consolidation is a time dependent phenomenon while compaction is an instantaneous process and faster than consolidation. Figure 27 shows compaction and consolidation in terms of phase diagram.

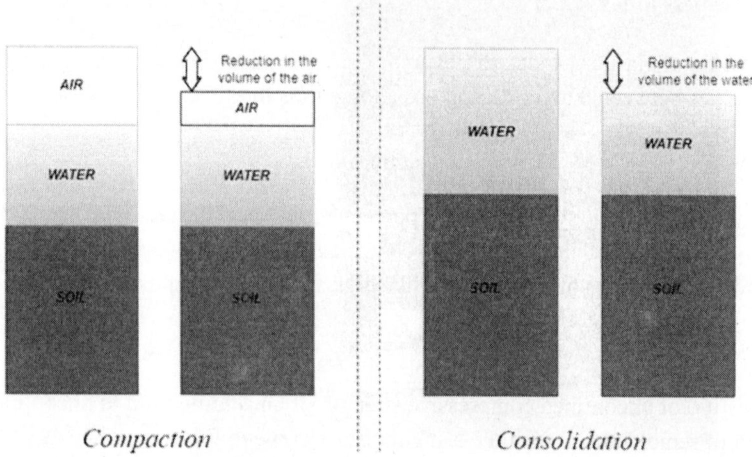

Figure 27: Phase diagram for compaction and consolidation.

(c) Permeability.

Permeability is the measure of how easily a fluid (water) can flow through a porous medium (soil). Permeability is maximum for coarse grained and minimum for fine grained soils. Permeability depends upon degree of saturation, void ratio, unit weight, viscosity, entrapped impurities, grain size etc. It is important for solving engineering problems such as seepage, piping in dams, yield problems in wells, settlement in hydraulic structures and designing dams structure, embankment, water retaining structure, culverts , pavement design etc.

3.1.9. Scope of soil mechanics

(a) Foundations

Any structural load must ultimately be transmitted to the soil through its foundation. Thus, understanding of soil mechanics is required to choose the type and details of a foundation. Foundations are essentially divided into two main types depending on the type of construction. Shallow foundation transfers the structural load very near to the surface while deep foundation transfers the load to subsurface layers. Generally, shallow foundations are used for structures with small to moderate loads, while deep foundations are used for tall buildings and megastructures.

(b) Underground structures

Underground structures (Figure 28) are built beneath the earth's surface and various types of forces such as earth pressure, hydrostatic force, seepage, etc. are exerted on them. They include metro, tunnels, water distribution pipes, and sewage networks. Design of such structures require complex geotechnical engineering knowledge.

Figure 28: Underground structure. (Source: http://www.quantity-takeoff.com)

(c) Pavement Design

Depending on the type of pavement, pavement design may be flexible or rigid. Rigid pavements distribute wheel loads over a wide area of the subgrade soil while flexible pavements distribute wheel loads to the lower layers of the pavement. The resistance to traffic loads for pavements is dependent on the subgrade soil, which is subjected to repetitive loading, swelling and shrinkage, and frost damage caused by thawing.

(d) Excavations, embankments, and dams

In order to design and construct embankments and earth dams efficiently, it is crucial to have an understanding of shear strength, seepage, stability of grade, compressibility, and subsequent uplifting and compaction. For example, excavations commonly require soil stability calculations, and deep excavations may require temporary support timbers or braces whose design requires an understanding of soil mechanics. Likewise, a knowledge of soil behavior in the presence of water is necessary for designing earth dams and embankments.

3.2. Rock Mechanics and Geology

Rock mechanics studies the characteristics of rocks and the unique methods used in the design and construction of engineering works built upon, through, and beneath the rocks. Civil engineers can benefit from an understanding of rock mechanics. Bridges, dams, and buildings built on rock foundations are examples of common civil engineering applications, as are numerous subterranean installations and tunnels, deep cuttings for spillways, stone quarries etc. In general, rocks are durable materials that can withstand far greater loads than soils. However, the strength of rock is limited. Whenever the loads are too high, rocks may fail. The mechanical characteristics of the rock material and geological discontinuities which include faults, joints, fissures, and so on influence the exact behavior of the rock layers when subjected to a change in stress. Bridges, dams, high buildings, long tunnels, and deep mines cause significant changes in the rock mass. A thorough understanding of the mechanical characteristics of rock masses as well as their behavior under various loading conditions is required for accurate assessment and correct prediction of rock behavior.

3.2.1. Rock

Rock is a substance consisting of mineral aggregates that are formed naturally. Rocks are broadly classified into three types: igneous, sedimentary, and metamorphic (Figure 29). Earth's external layer of solid material is composed of rock. Petrology is the study of rocks.

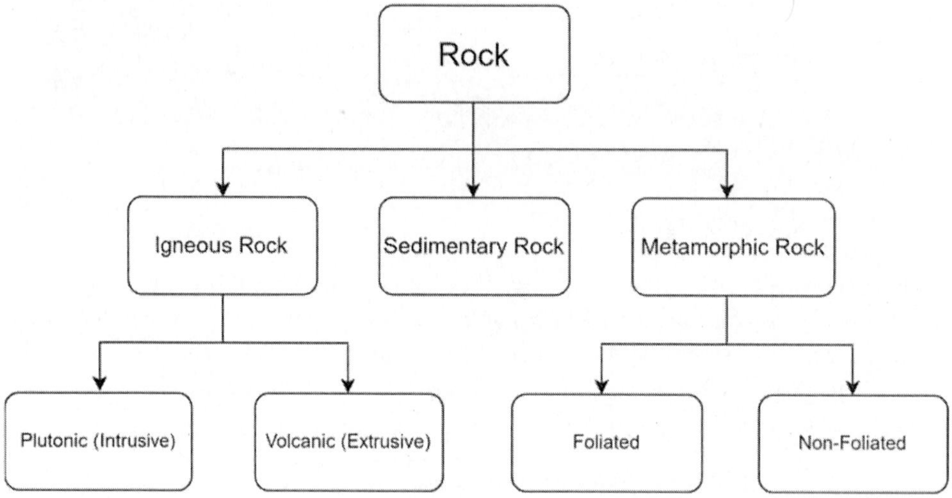

Figure 29: Classification of rocks

(a) Igneous rocks

These rocks occur due to the solidification of lava or magma beneath the earth. They are crystalline in nature and are further classified as volcanic and plutonic. Examples of igneous rocks are basalt, trap, granite, diorite, pegmatite etc. as shown in Figure 30. Volcanic rocks are fine grained and extrusive in nature as they are formed when magma erupts and comes in contact with external or outer surfaces. Plutonic rocks are intrusive in nature as they form when magma cools beneath or under earth. They are coarse grained as they cool down slowly which results in large crystal growth.

(a) (b) (c)

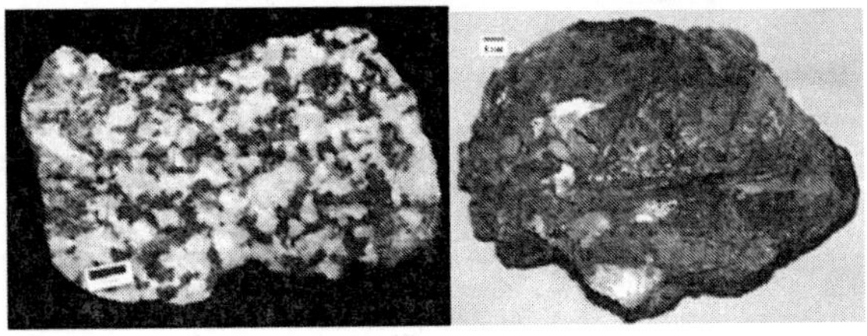

(d) (e)

Figure 30: (a) Basalt (b) Trap (c) Granite (d) Diorite (e) Pegmatite (Source: Wikipedia)

(b) Sedimentary rocks

These rocks are formed due to weathering of existing rocks through physical or chemical processes. Weathered material is transported by wind, water, gravity, etc., consolidated, and compacted in layers. Examples of sedimentary rocks are sandstone, limestone, shale, gypsum, laterite etc. as shown in Figure 31.

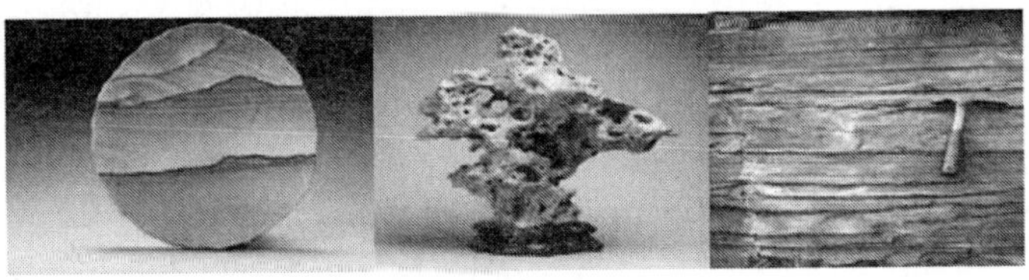

(a) (b) (c)

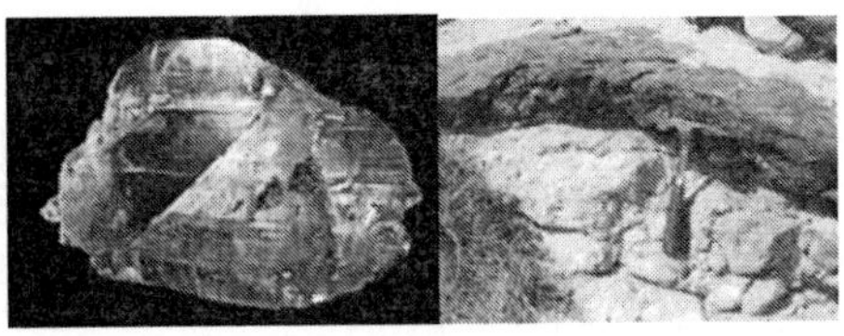

(d) (e)

Figure 31: (a) Sandstone (b) Limestone (c) Shale (d) Gypsum (e) Laterite (Source: Wikipedia)

(c) Metamorphic rocks

When pre-existing rocks such as igneous and sedimentary rocks are subjected to high temperature and pressure conditions, their composition and character changes to hard and durable material. Examples are marble, quartz, gneiss etc., as shown in Figure 32.

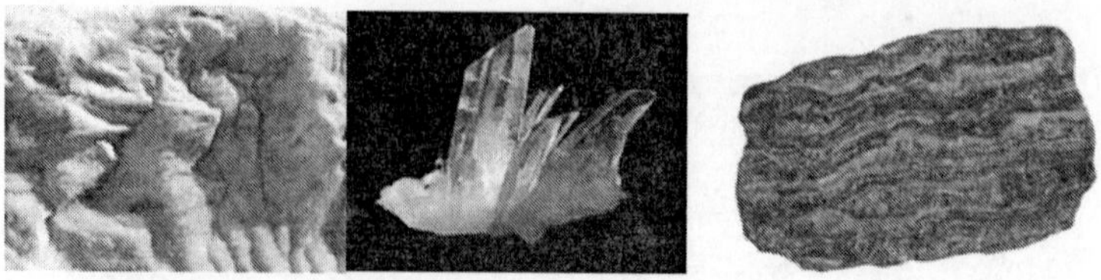

(b) (c)

Figure 32: (a) Marble (b) Quartz (c) Gneiss (Source: Wikipedia)

Features of rocks

(a) **Texture:** The geological term "texture" refers to the arrangement, distribution, and shape of minerals or grains within a rock. The term texture is not synonymous with the roughness of the rock surface.

(b) **Structure:** Structure refers to the broader characteristics of a rock that extend such as bedding (in sedimentary materials) and foliation (in metamorphic materials).

(c) **Grain size:** Grain size refers to the average diameter of the individual particles of sediments and rocks. It is important because it conveys important information on the historical conditions of erosion and transportation.

(d) **Mineralogy:** Mineralogy discusses the type of minerals present in rock in different proportions, for example quartz consists of feldspar, quartz, and foliate.

(e) Various other parameters such as hardness, color, strength, and specific gravity determine rock physical properties.

Modes of failure of rocks

(a) **Flexure failure:** Due to bending action on rock, tensile cracks are formed which propagate due to its own weight. As sag increases, crack propagation increases resulting in failure of rock.

(b) **Shear failure:** Rupture on the surface of rock is formed as shear stress exceeds its critical value. Due to rupture, the expulsion of shear stresses causes the displacement of rocks along the rupture (Figure 15).

(c) **Direct tension failure:** This type of failure generally occurs in rock resting on a sloped surface subject to direct tension (Figure 15). Both gravity and friction are applied on an inclined sloped surface resulting in increase of pull and tension failure.

(d) **Crushing or compression failure:** This type of failure occurs when compressive stresses exceed its critical value resulting in crushing failure (Figure 33).

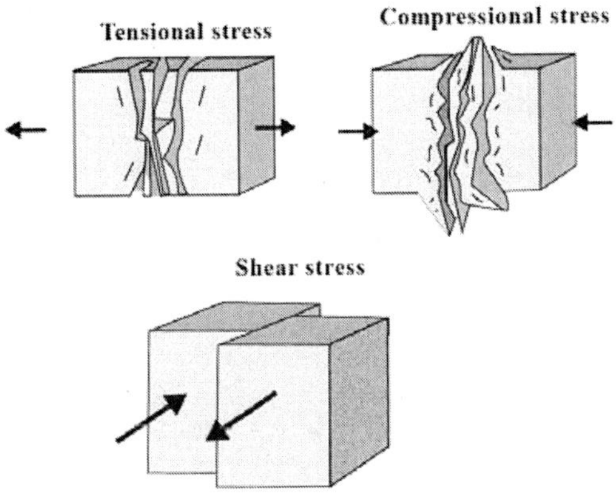

Figure 33: Rock failure types. (Source: Michael Kimberly, https://earthquake.usgs.gov/)

3.2.2. Importance of engineering geology

Sites with a large presence of faults and folds pose a significant risk to engineering structures. The presence of faults or breaks reduces the load bearing capacity of the rocks. The knowledge of engineering geology is required to avoid the high costs that may be borne by the concerned authorities in the event that they set up these projects on the areas of joints and faults and folds, as well as the lives that may be at risk.

3.2.2.1. Structural geology

Geology of structural forms studies the origins, occurrences, classifications, types, the consequences of secondary structures, such as folds, faults, joints, and rock cleavages, which are distinct from primary structures, such as bedding and fasciculations, which are developed in the formation of rocks. Some of the structural elements are described below:

(a) **Strike**: A geological structure is referred to as striking in a particular direction as shown in Figure 34. Spike direction is defined as the direction of the curve at the intersection of the bedding planes.

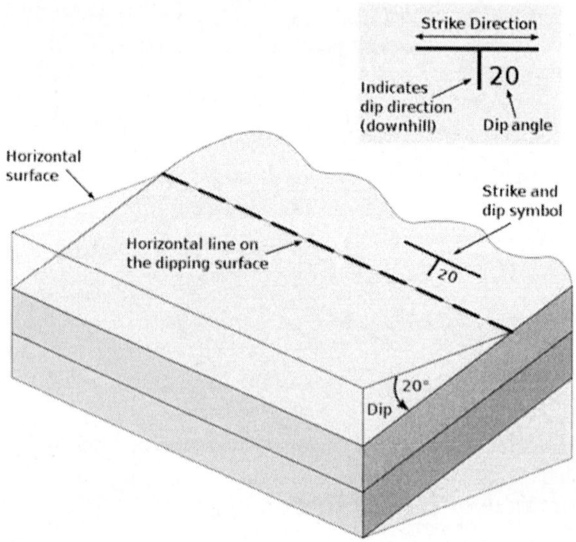

Figure 34: Strike & Dip. (Source: Physical Geology Book, Dr. Karla Panchuk, 2018)

(b) **Unconformity:** When sedimentary rocks are significantly different in age, or when sedimentary rocks have been eroded and have been exposed to older igneous or metamorphic rocks, an unconformity occurs. These unconformities represent periods of time without rocks to represent them.

(c) **Joint:** A joint, as shown in Figure 35, is a fracture separating two unmoving sections of a rock. There is no displacement in the joints unlike faults.

Figure 35: Joint. (Source: HylgeriaK, 2010)

(d) **Folds:** Geological structures known as folds are common in rocks (Figure 36). When the horizontal layers are compressed, they bend either upward or downwards. Folds are ductile deformations caused by rocks bending under stress. The folded sides are called limbs, and the hinge zone is the region of curvature where the limbs meet. Folds have an axial surface whose main function is to divide the fold into two halves. The axial trace is the line formed between the fold's axial surface and another surface, such as the bed's top.

Figure 36: Parts of folds (Source: Physical Geology Book, Dr. Karla Panchuk, 2018)

(e) **Fault:** The term fault describes the movement of one side of a surface or narrow zone relative to the other side in a direction parallel to the surface or narrow zone. Most faults are brittle shear fractures or zones of closed shear fractures, but some are narrow shear zones in which ductile deformation occurs without loss of cohesion at the outcrop scale. Faults basically occur due to shear fracture that propagates to certain meters and affects the block of earth, major plate boundaries hundreds or even thousands of kilometers long. There are two types of faults - strike-slip and dip-slip. Strike–slip faults (Figure 37) are primarily horizontal displacements of rocks along faults and can be further classified into dextral and Sinistral depending on orientation. A dip-slip fault (Figure 38) is an inclined rupture in which the blocks frequently shift up and down. They can be further classified into normal or reverse faults depending on the angle and relative movement of the hanging wall and footwall.

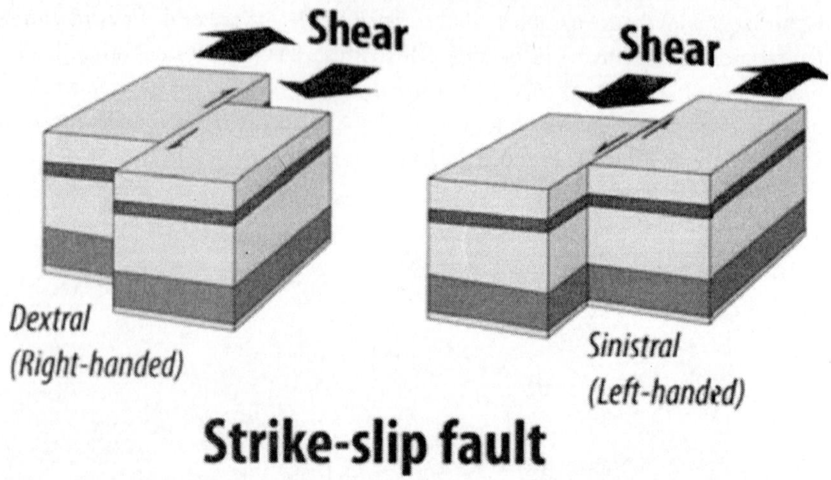

Figure 37: Strike-slip fault (Source: Physical Geology Book, Dr. Karla Panchuk, 2018)

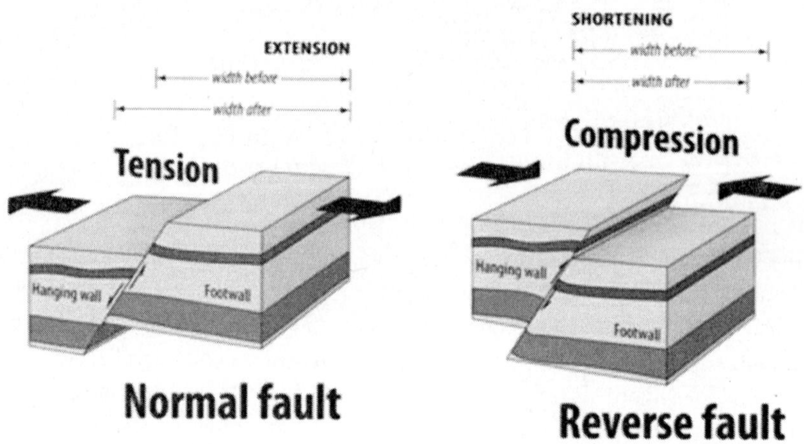

Figure 38: Dip-slip fault (Source: Physical Geology Book, Dr. Karla Panchuk, 2018)

3.3. Types of Foundations

Foundation is the part of a civil engineering structure which transfers the superstructure load to ground and is in direct contact with the ground. It acts as the linkage between building structure and ground. A well-designed foundation is crucial for longer serviceability of a structure. Foundations are divided into two main types depending on the type of construction. Generally, for structures with small to moderate loads, shallow foundations are used, and deep foundations are adopted for tall buildings and megastructures. A foundation can also be classified according to the type of footing. There are both on-ground and underground foundations, each with its own construction methods and techniques. For building foundations, it is necessary to conduct a detailed ground and soil study, as it dictates the behavior of the structure, which materials are to be used, and whether the soil can bear the load coming from the structure. The weight of superstructure above the ground level is distributed over sufficiently large areas to avoid uneven settlement. The general requirements for foundation design are:

(a) **Shear failure criteria or bearing capacity:** Foundation must be safe against shear failure.

(b) **Settlement criteria:** Settlement must be within a safe limit, as differential settlement may happen due to unequal load changes.

(c) **Location & depth criteria:** Foundation performance should not be affected by seasonal volume changes of soil due to swelling and shrinkage and also by the presence of adjoining structures.

3.3.1. Classification of foundation

Karl Von Terzaghi was an Austrian geotechnical engineer who provided a classification of foundations based on the ratio of depth of footing to width.

3.3.1.1. Shallow foundation

If the ratio of depth of footing to width is less than or equal to 1 then the foundation is called shallow foundation (Figure 39). Shallow foundation transfers the load at a small depth. Examples of such foundations include combined footing, raft footing, isolated footing etc. The shallow foundation is placed or located below the column to transfer and distribute wall and column load by end bearing mechanism. As the footing load spreads, the bearing pressure decreases with depth and the soil which is in direct contact of footing takes the maximum load. This type of foundation is preferred for small and moderate-sized houses and are comparatively cheaper and easy to construct for domestic houses.

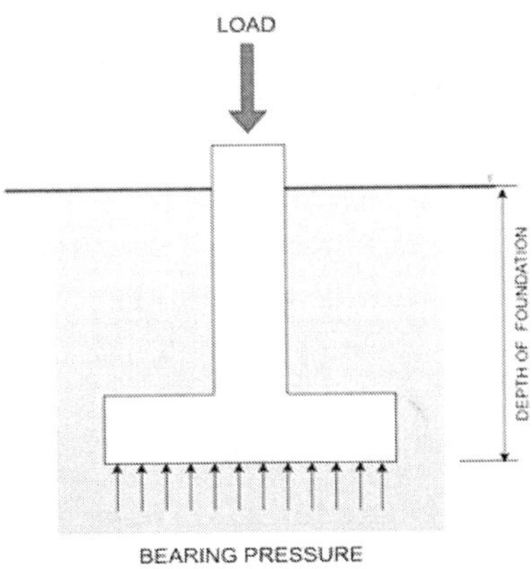

Figure 39: Shallow foundation (Source: Project Lead the Way, PLTW)

3.3.1.2. Deep foundation

When the ratio of depth of footing to width of footing is more than 1 (generally, 4-5), then the foundation is called deep foundation. An example of this is pile foundation. Deep foundation transfers the load of the structure vertically from the superstructure to the deeply buried rock strata. It uses end bearing and skin friction mechanism to transfer the structure load to rock strata at a greater depth as shown in Figure 40. Deep foundations are used when shallow foundations are inadequate to transfer the structural load to the ground surface.

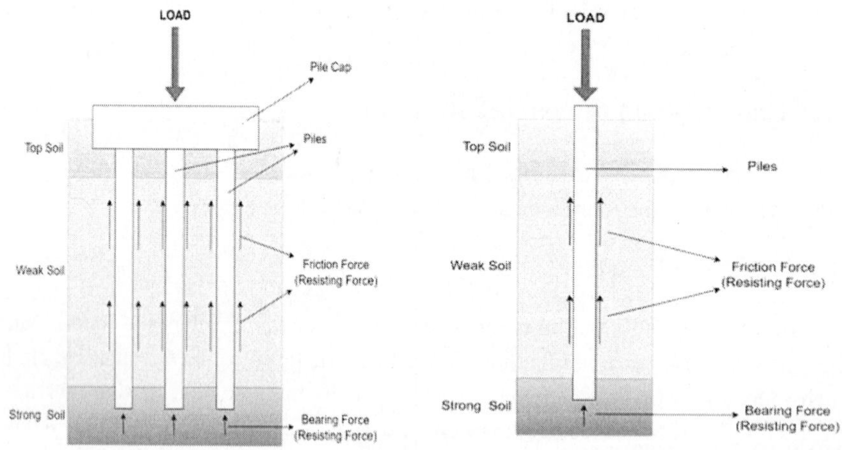

Figure 40: Deep foundation (Source: PLTW)

Piles are designed to transmit structural loads through low density soil zones to a depth at which soils have the capacity to sustain the loads. Pile foundations are typically used when soil beneath the foundation lacks the bearing capacity to support the structural load into subsoil up to bedrock. Piles are columns with a small diameter that are driven or cast by creating a hole and pouring concrete into the ground. They are made of materials such as cement, wood, steel, or composites. Pile foundations are used when the top stratum of soil is highly compressible and weak strength. These types of foundations are usually employed to build bridges (Figure 41).

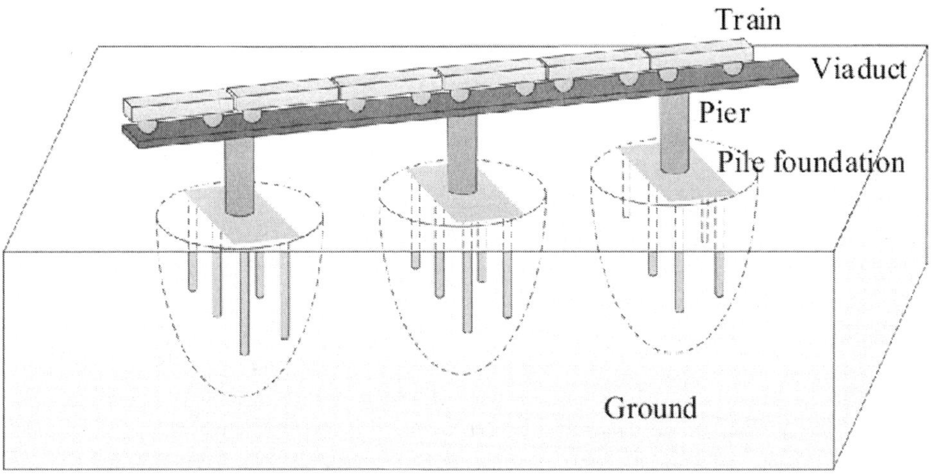

Figure 41: Pile foundations in bridges. (Source: Wu et al. (2002))

3.3.2. Types of shear failure

By investigating the failure patterns of foundations, Terzaghi classified the shear failure of soil under a foundation base into general and local shear failure, and then Vesic added the third type based on the type of soil and the location of the foundation.

(a) General shear failure

- Occurs in shallow foundations when placed on medium to dense soil.
- When a foundation fails, it becomes slanted, and heaving occurs on the sides.
- Small settlements occur before they collapse, and the stress zone continues up to the ground level.
- Clear failure point is obtained in load v/s settlement curve.

(b) Local shear failure

- Occurs in shallow foundations, when placed on loose sand or soft clay.
- There may be little to no heaving at the sides.
- Before failure, large settlement occurs, and the stress zone doesn't extend up to ground level.
- Progressive failure is obtained in load vs. settlement curve.

(c) Punching shear failure

- Occurs in deep footing or piles which are placed on loose or soft soils.

• Soil below the foundation gets cut off from the adjacent soil by shearing and excessive settlement is recorded.

• The adjacent soil remains undisturbed and there is no tilting and heaving at the sides.

3.4. Underground Structures

Underground constructions serve a variety of purposes such as railways, subways, motorways, sewage networks, and water transport. Tunnels, caverns, shafts, and their components are examples of underground structures that are totally enclosed and contained within the existing host ground medium.

Considering the relationship of an underground structure with its host medium is crucial for a safe and effective design. Once excavation is carried out for the construction of an underground structure, there comes a period of adjustment in the particles of subterranean media. This length of time is required for the soil to return to stability and equilibrium. The properties of subsurface media change during the period of adjustment and must be accounted for in the design. Moreover, while constructing any underground structure, especially tunnels, there is a lack of space which makes it difficult to haul material from one place to another. The construction, and instrumentation of an underground structure requires more careful planning, design, and construction sequencing than those for surface and aerial structures.

3.4.1. Types of underground structures

(a) Tunnels

A tunnel is a completely contained underground corridor built through the surrounding dirt, earth, or rock, with the exception of the entrance and the exit ports. It can be utilized for pedestrian and vehicular traffic, as well as rail and canal transportation. Tunnels are frequently used as part of a rapid transportation system. Some special purpose tunnels such as wildlife-crossing tunnels are designed to help wildlife safely navigate human-made barriers such as highways. An important tunnel in India is the Pir Panjal railway tunnel in Jammu and Kashmir, the longest railway tunnel in India with a length of 11.215 km and located at an altitude of 36795 ft. Another strategically important tunnel currently being constructed is the Sela Tunnel in Arunachal Pradesh, which is proposed to enable all-weather connectivity between Guwahati in Assam and Tawang in Arunachal Pradesh.

(b) Shafts

The shaft is a vertical or inclined tunnel that extends from the surface of the earth to the roof of the tunnel. A shaft provides a wide range of advantages and uses in tunnel construction. Shafts are dug at appropriate places along the tunnel line to suitably facilitate tunneling operations when the course and position of the tunnel is clearly outlined.

(c) Caverns

A cavern is a subterranean opening with an axial dimension of not more than fifteen times the lateral dimension. Caverns are used for underground pumped storage, power houses, power plants, subway stations, and storage facilities. The ideal shape of the roof of a cavern is a circular arch, however shapes like multi radial or elliptical can also be used depending on the suitability with geology of the sub-strata.

3.4.2. Surrounding media

The host medium has an active and substantial influence in the structural behavior of a subterranean construction. This necessitates considering the constitutive connection of the host while designing the subsurface structure. However, determining ground features in a quantitative manner is challenging. The challenge arises from the fact that the ground is often heterogeneous or stratified, and hence its features cannot be anticipated using pointwise drill hole logging or other subsurface approaches. Although there are various subsurface exploration techniques available, one must be able to answer some key questions:

(a) What is the nature of media present beneath the surface of the ground (soft/rocky)?

(b) Is the host media stable enough?

(c) What is the depth of the water table at the location concerned?

(d) Is host media susceptible to squeezing and swelling?

(e) If the host media is rocky in nature, then does the location of our structure fall in the rock burst zone?

This list of questions is certainly not exhaustive and should include other relevant queries regarding the characteristics of subsurface media.

3.4.3. Use cases of underground structures

It is important that structures-built underground is durable enough to withstand adverse climatic conditions such as hurricane winds, sudden drops in temperature, mudslides, long downpours, or flooding in an area with sloping terrain. In the ore-mining business, subterranean structure development has advanced dramatically. In addition to hydraulic engineering and transport tunnels, underground structures can include metro stations, hydroelectric power plants, cold storage, underpasses for pedestrians, garages and municipal supply systems, water storage reservoirs, warehouses, oil and gas storage installations, containers for the dumping of hazardous waste, industrial projects, military installations, and medical facilities. Below are some areas of application of underground structures:

(a) **Underground power plants:** Due to advantages such as reduced length of supply mains, requirement of lesser concrete work, and overall lesser amount of materials consumed; underground power plants are more cost effective to operate than surface power plants.

Approximately Several million cubic meters of rock are excavated when a huge subterranean hydroelectric power station is constructed. For Example:

•Ichari Dam: The first subterranean power plant in Northern India was the concrete gravity dam erected across the Tons River in Dakpathar in Uttarakhand. The concrete gravity and run-of-the-river dam are 59 meters tall and 155 meters long. Water from the reservoir created by the dam is diverted to the Chibro powerhouse which has a capacity of 240MW.

•The Nathpa Jhakri Hydroelectric Power Plant is the country's largest underground hydropower plant having a production capacity of 1500 MW.

(b) **Underground warehouses:** In addition to their flexibility to accommodate current mining operations beneath them, these structures provide stability, fire resistance, defy humidity, and temperature changes. As limited entry and exit points are provided in the case of underground warehouses, security can be tightly controlled. As space above the ground is freed up, it can be utilized for other purposes. Underground warehouses can be used for either live or dead storage. On a daily basis, live storage involves the movement of a high number of items and materials. Thus, well-designed loading docks and direct rail connections between warehouses and train terminals are required for live storage.

(c) **Urban Infrastructure:** In large cities, the areas underneath cities are undergoing massive development, especially in the transportation sector. In already overcrowded cities, land availability, even for government-funded infrastructure projects, is hard to come by and the authorities are aggressively pushing for developing underground structures. This helps to optimize areas above earth's surface for maximum efficiency, administer transportation services, decrease traffic mishaps and accidents, enhance urban aesthetics, and reduce vehicular air and noise pollution. Several types of urban underground structures that fall under the transportation domain include pedestrian tunnels, vehicular tunnels, parking lots, and railroad terminals. In the service sphere, underground establishments include a variety of commercial establishments such as cafés, stores, cinema halls, cold storage facilities, exhibition halls, and archives.

(d) **Underground Storage:** Petroleum products and, sometimes, potable water is stored underground, where they can be stored in reservoirs with capacity of several million cubic meters. Such high capacity is difficult to achieve in the case of elevated or surface storages simply because land is at a premium. RCC, prestressed, and prefabricated panels can be used in the construction of underground storage reservoirs. An example of underground storages is the "Strategic Petroleum Reserve" which has become the cornerstone of the geopolitical strategy of many countries to ensure economic stability during volatile periods. Countries like India in recent times have expanded their "Strategic Oil Reserves," which currently stands at 5.33 million Tonnes. Indian Strategic Petroleum Reserve (ISPRL) presently operates 3 strategic reserves at Padur (2.5 MMT), Visakhapatnam (1.33 MMT), and Mangalore (1.5 MMT). During the initial days of the pandemic and lockdowns when global oil prices were tumbling down to historic lows, India purchased large quantities of oil and replenished the strategic oil reserves to safeguard their energy security. For storing petroleum products, usually underground storage is preferred because evaporation losses are reduced, and it is easy to discharge oil into caverns directly from ships.

3.4.4. Future Perspective

Over the last few years, tunneling has been utilized for a range of purposes. With ever increasing urbanization, the cities will have no option but to explore innovative ways to accommodate the burgeoning population. Due to limits in surface development and in addition to ecological concerns, the future population will require considerable utilization of underground construction. Moreover, underground construction provides natural protection to whatever is buried beneath it. It provides the advantage of shielding the surface environment from the dangers and disruptions that come with particular operations.

3.5. SOLID WASTE MANAGEMENT AND LANDFILL

The practice of managing and collecting solid wastes is termed as solid waste management. Inappropriate disposal of municipal solid waste results in unhygienic conditions and has a large impact on public health. A proper waste management system is crucial for maintaining social hygiene.

3.5.1. Categories of Waste

- Organic waste: Kitchen waste, fruits, vegetables.
- Ashes or Dust: It is the residue left after burning fuel for cooking.
- Construction waste: Wasted concrete, roofing, etc.
- Hazardous waste: Industrial waste, medicinal wastes.
- Dead animals: Carcasses of dead animals.
- Soiled waste: Hospital waste such as cloth soiled with body fluids.
- Recyclables: Paper, glass, metals, plastics.
- Toxic waste: bulbs, spray cans, batteries, chemicals, thinner and paint, containers, fertilizer, and pesticide.
- Bulky waste: Tires, tree branches, etc.
- Combustibles: Dried leaves, Paper, wood, packaging for relief items etc.
- Non-combustibles: Bottles, stones, metal, tins, cans etc.

3.5.2. Various Sources of Solid Waste

(a) **Residential:** Human dwellings are major sources of solid waste. Most houses dispose of their solid wastes in garbage cans, which are subsequently emptied by a rubbish collection business or municipality for treatment.

(b) **Industrial:** The industrial sector is a huge solid waste contributor. They include heavy and light manufacturing firms, construction sites, fabrication factories, canning plants, chemical plants, power plants etc. These industries produce packaging wastes, food wastes, ashes, materials from demolition of construction sites and other hazardous wastes.

(c) **Construction & Demolition Areas:** Construction and demolition sites generate a variety of solid wastes in the form concrete, steel, rubber, dirt, glass, and plastic.

(d) **Agricultural Waste:** Crop farms, vineyards, dairies etc. also generate solid waste. The waste they generate include spoiled food, agricultural debris, and toxic materials.

3.5.3. Methods for the Management of Solid Waste

There are various strategies for managing solid waste. The following are examples of acknowledged techniques:

(a) Landfill

A municipal solid waste landfill is a designated land area or excavation that receives waste. Essentially, garbage is thrown out in thin layers, compacted, and covered with plastic foam or dirt. Present era landfills are constructed so that the bottom is lined with an impermeable liner, which often consists of multiple layers of sand and thick plastic. This liner plugs the groundwater contamination by percolation. When the landfill is filled, layers of topsoil and gravel together with sand and clay are applied to prevent water seepage.

(b) Incineration

This practice involves the combustion of solid wastes at high temperatures until they are reduced to ash. The design of incinerators prevents them from emitting excessive heat when burning solid trash. Waste-to-energy plants are incinerators that recycle thermal energy through furnaces and boilers. These waste-to-energy systems are costlier to install and maintain than conventional incinerators because they require specialized equipment and controls, highly trained technical employees, and auxiliary fuel supplies. The advantage of this procedure is that it reduces waste volume by up to 20-30% of the initial volume. However, the disadvantage is that in addition to posing a fire or smoke concern, it also produces gaseous contaminants.

(c) Recycling and Recovery

Recycling or resource recovery is the practice of reusing objects that are still useful but have been discarded. The most industrialized nations have a strong heritage of recycling to reduce trash levels. It is an environmentally friendly method of waste management, though requires extensive civic participation and investment in infrastructure.

(d) Decomposition

Due to a lack of enough room for landfills, biodegradable waste is allowed to decompose in a specially controlled environment. In this biological process, microorganisms, mainly bacteria and fungi, transform biodegradable organic waste into humus-like compounds. This substance, which has the appearance of soil, is rich in carbon and nitrogen. Compost produced is rich in minerals and perfect for plant growth. However, for large-scale operations, rigorous management and skilled employees are required.

(e) Pyrolysis

In this practice, solid wastes are chemically destroyed by high temperature in the absence of oxygen. The temperature goes up to 430 degrees Celsius and, under pressure, solid wastes are converted into gases, solid carbon and ash residue, and minor amounts of liquid. However, the systems that degrade chlorinated organic molecules with heat may produce incomplete combustion byproducts such as furans and dioxins. These hazardous chemicals are produced at parts per trillion concentrations which require proper treatment, storage, and disposal.

Geotechnical engineering is a dynamic sub-field of civil engineering involving thorough investigation and understanding of the behavior of soil and rocks under various physical conditions. Geotechnical engineering substantially influences the sustainability of any infrastructure and hence a solid knowledge of the subject is essential for the design and development of any civil engineering structure.

References and suggested readings

1. Das, Braja M.., Sobhan, Khaled. Principles of Geotechnical Engineering. United States: Cengage Learning, 2013.

2. Das, Braja M, Sivakugan, Nagaratnam. Introduction to Geotechnical Engineering. United States: Cengage Learning, 2015.

3. Ranjan, Gopal. Basic and Applied Soil Mechanics. India: New Age International (P) Limited, 2007.

4. Ghulati, Sahashi K and Datta, Manoj. Geotechnical Engineering. McGraw-Hill Education (India) Pvt Limited, 2005.

5. Indian geotechnical society annual lectures. (http://www.igs.org.in:8080/portal/igs-publications/igs-annual-lecture.html)

6. Indian standard codes for soil and foundation engineering. (https://law.resource.org/pub/in/bis/manifest.ced.43.html)

7. Wu, Y., Bian, X., Cheng, C., Jiang, J., 2022. A substructure approach for analyzing pile foundation and soil vibrations due to train running over viaduct and its validation. Rail. Eng. Science. https://doi.org/10.1007/s40534-022-00276-z

Dynamic QR Codes for Further Reading

1. USDA Soil Texture Calculator. Source: https://www.nrcs.usda.gov/	Soil Texture
2. What is Geotechnical Engineering? Source: The International Society of Soil Mechanics and Geotechnical Engineering (ISSMGE)	Geotechnical
3. International Society for Soil Mechanics and Geotechnical Engineering Virtual University Source: http://virtualuniversity.issmge.org/	ISSMGE VU
4. Why do buildings need foundations? Source: Practical Engineering	Foundations

4 Water Resources Engineering and Sustainability

Unit specifics

Through this unit we have discussed the following aspects

- *Introduction to water cycle and fluid mechanics*
- *Basic concepts of wave and current system*
- *Basic concepts of sediment transportation*
- *Introduction to water resource structures such as multipurpose reservoirs*
- *Introduction to wastewater and effluent treatment systems*
- *Introduction to water supply systems*

The practical applications of the topics are discussed for generating further curiosity and creativity as well as improving problem solving capacity.

After the related practical, based on the content, there is a "Know More" section. This section has been carefully designed so that the supplementary information provided in this part becomes beneficial for the users of the book. This section mainly highlights the initial activity, examples of some interesting facts, analogy, history of the development of the subject focusing the salient observations and finding, timelines starting from the development of the concerned topics up to the recent time, applications of the subject matter for our day-to-day real life or/and industrial applications on variety of aspects, case study related to environmental, sustainability, social and ethical issues whichever applicable, and finally inquisitiveness and curiosity topics of the unit.

Rationale

This fundamental unit on water resource engineering and sustainability gives a basic idea about the various processes and concepts involved in water resource engineering along with water treatment and distribution systems. This chapter starts with a discussion and explanation of the fundamentals of wave and current systems along with sediment transport systems. It then discusses the various types of water resource structures such as multipurpose reservoirs, dams etc. The last section of the chapter discusses wastewater treatment processes and water distribution systems. This chapter discusses the topics to a length with practical examples such that it can develop the preliminary ideas and concepts of the subject.

Water resource engineering deals with the qualitative and quantitative study of the various water resources systems. This subject covers a large band of topics starting with the fundamental process of water resources such as water cycle, study of flow of water, engineering structures of water resources, wastewater treatment, distribution systems of water etc. Water resource engineering is the branch of civil engineering that studies the methods to manage water with scientific and engineering techniques.

Pre-requisites

Mathematics: Coordinate Systems (Class XII)

Physics: Mechanics (Class XII)

Unit outcomes

List of outcomes of this unit is as follows:

U1-O1: Understand basic concepts of water cycle and fluid mechanics.

U1-O2: Understand the various types of structures used to manage water resources.

U1-O3: Understand the basic concepts and components of wastewater treatment systems and water distribution systems

Water covers about 71% of the earth's surface and there is roughly 326 million cubic miles of water on the planet. Only 0.5% of this is in the form of available freshwater. Most of this water is in groundwater, less than 1/2 mile deep within the earth and rest is in freshwater lakes and streams. Because of the diversity in the quality and quantity of the amounts of water involved, and the myriad use cases, civil engineers must deal with a multitude of physical and management problems. Water resources engineering can be broadly subdivided into hydrology and hydraulics. Hydraulics is the study of motion of water that emphasizes the mechanics of water flow, open channel flow, and flow-structure interactions. Hydrology is primarily associated with watershed and river modeling and understanding interactions between atmospheric, surface, and subsurface water.

4.1. Fluid Properties

Fluids are substances that can flow and do not conform to a fixed shape. If any shear stress is applied, it can deform continuously. Fluids can be put into two groups: liquids and gasses. The main difference between liquids and gasses is that liquids are virtually incompressible, while gasses can be compressed and usually need to be. Another difference is that liquids have a set volume and free surfaces, while gas expands until it fills up the whole container. Fluid mechanics and hydraulics is the study of how fluids behave when they are at rest and when they are in motion. Kinematics is the science that studies the geometry of fluid motion without reference to the forces causing the motion. Consequently, kinematics is concerned with describing the motion of fluids in terms of their space-time relationship. Kinetics, on the other hand, is the science concerned with the action of forces in generating or altering the motion of fluids. Clearly, the study of fluid motion must take both kinematics and kinetics into account.

There are two ways to describe the motion of a fluid: the Lagrangian method and the Eulerian method. In the Lagrangian method, a single fluid particle is chosen and followed throughout its entire path of motion, and the behavior of this particle during its path of motion through space is observed. In the Eulerian method, any point within the space occupied by the fluid is chosen, and any changes in velocity, density, and pressure at that point are observed.

A fluid is made up of distinct particles that move at varying speeds and may be susceptible to varying accelerations. In addition, the velocity and acceleration of a fluid particle may vary in both time and space. Therefore, it is essential to examine the movement of fluid particles at numerous sites in space and at successive moments in time while studying fluid flow. In fluid flow, there are three important ideas. The first is the principle of conservation of mass, which is where the equation of continuity comes from. Then the principle of kinetic energy, from which some flow equations are made, and finally, the principle of momentum, from which equations can be made to measure the dynamic forces that fluids exert as they move.

4.1.1. Types of Fluid Flow

Fluid flows can be categorized in a number of ways based on various factors:

(a) Steady and unsteady flow

(b) Uniform and non-uniform flow

(c) Laminar flow and turbulent flow

(d) One, two, and three-dimensional flow

(e) Rotational flow and irrotational flow

Let us understand what these flows represent.

(a) **Steady and unsteady flow**

Fluid flow is considered to be steady if, at any point in the fluid, the parameters that describe the behavior of the fluid in motion, such as velocity, temperature, pressure, and density, do not vary with time. In other words, a steady flow is one in which, at any point in the moving fluid, different characteristics that describe how the fluid is moving are the same over time.

Unsteady fluid flow occurs when, at any point in the fluid, one or more of the features that describe the behavior of the fluid in motion vary with time. Thus, a fluid flow is unstable if, at any point, the fluid is slowing It could mean, though, that when it comes to vector quantities like flow velocity, even a change in the direction of such quantities with respect to time at any point in a flowing fluid can make the fluid unsteady.

Stable flow is easier to evaluate than unsteady flow. Moreover, most engineering problems involve steady flow conditions. Figure 42 illustrates steady v/s unsteady flow conditions.

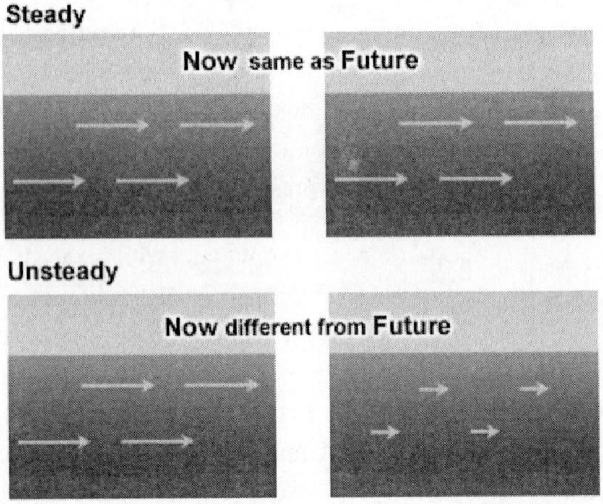

Figure 42: Steady vs. Non-Steady Flow (Source: MetEd UCAR)

(b) Uniform and non-uniform flow

Flow is said to be uniform when the speed and direction of the flow of a fluid doesn't change from one point to another in the flow at any given moment. At any instant in time, there is no change in the velocity vector in any direction throughout the whole fluid flow. Flow of liquids under pressure via lengthy pipelines with constant diameter is an illustration of uniform flow. On the other hand, in non-uniform flow, the velocity of the fluid varies from point to point within the flow at any given instant. Figure 43 shows uniform flow and non-uniform flow. An example of uniform flow is a consistent flow along a long, straight pipe with a fixed diameter. A non-uniform flow is demonstrated by steady flow through a pipe with bends or a pipe with a varying diameter.

Uniform vs. Non-Uniform Flow

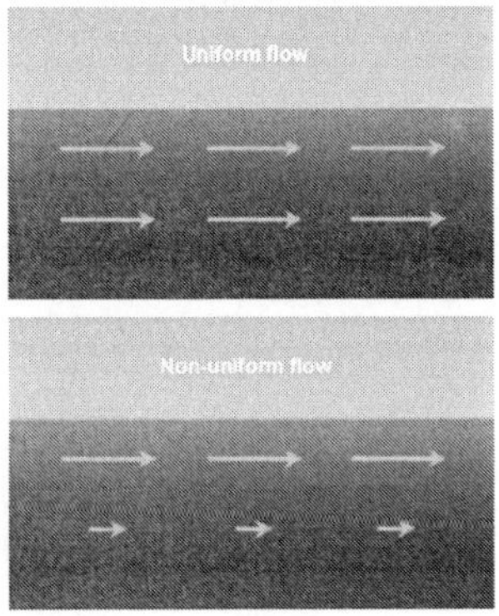

Figure 43: Uniform and non-uniform flow (Source: MetEd UCAR)

(c) Laminar flow and turbulent flow

When different fluid particles move in layers (or laminae), with one layer of fluid sliding smoothly over an adjacent layer, the flow is said to be laminar. The viscosity of the flowing fluid is crucial in the formation of a laminar flow. As a result, laminar flow may generally be used to describe the flow of an extremely viscous fluid.

A turbulent flow is one in which fluid particles move in a zigzag pattern. Due to the flow of fluid particles in a zigzag pattern, eddies arise, which result in significant energy loss.

The Reynolds number, a non-dimensional number, is used to categorize flows in pipes. The flow is referred to as laminar if the Reynold number is less than 2000. The flow is referred to as turbulent if the Reynold number is higher than 4000. Depending on whether the Reynolds number is between 2000 and 4000, the flow can be turbulent or laminar. Figure 44 illustrates laminar and turbulent flow.

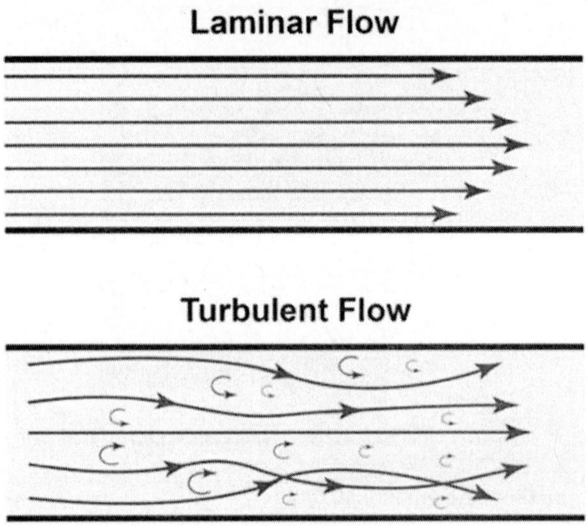

Figure 44: Laminar and Turbulent flow. (Source: Schetz and Allen E Fuhs, 1999)

(d) **One, two, and three-dimensional flow**

One dimensional flow: One-dimensional flow is a term used to describe a flow in which properties of the fluid, such as velocity, pressure, density, temperature, etc., depend only on one of the three coordinate directions and time t. Similar to this, a steady one-dimensional flow will result if the properties of the fluid are constant across time.

Two-dimensional flow: Two-dimensional flow occurs when the various properties of a flowing fluid are functions of just any two of the three coordinate directions, and time t.

Three-dimensional flow: The different properties of a fluid in motion are often functions of space and time; that is, they can change depending on the coordinates of any point (x, y, and z), as well as the passage of time (t). Three-dimensional flow is the name given to such a flow. It will be a steady three-dimensional flow if any of these properties of a flowing fluid do not change over time. Figure 45 illustrates the one dimensional, two dimensional and three-dimensional flow.

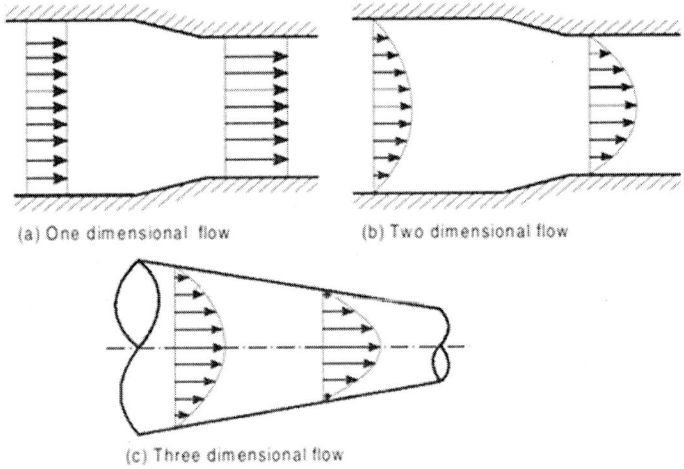

Figure 45: One dimensional flow, two dimensional and three-dimensional flow. (Source: Hydraulics & Fluid Mechanics including Hydraulics Machines)

(e) Rotational and irrotational flow

If fluid particles rotate around their mass centers while traveling in the direction of the flow, the flow is said to be rotational. The flow is said to be irrotational if the fluid particles don't spin around their mass centers while moving in the direction of flow. However, it may be claimed that a real irrotational flow exists only for flows of perfect fluids in which there are no tangential or shear stresses. Nevertheless, the flow of practical fluids may be assumed to be irrotational if the viscosity of the fluid is insignificant. Figure 46 illustrates rotational and irrotational flow.

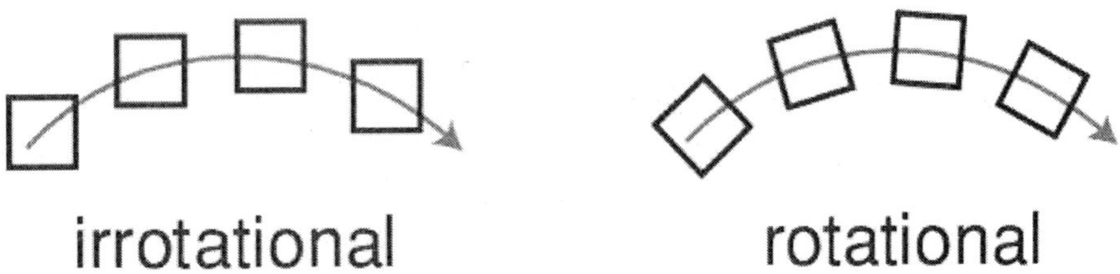

Figure 46: Rotational flow and irrotational flow (Source: Eeltink et al. 2020)

4.2. Water cycle and fluid mechanics

Water cycle is the vertical and horizontal movement of water through earth and atmosphere through various mechanisms such as evaporation from land and ocean, rainfall and snowfall, groundwater movement, and streamflow. Water evaporates from the sea and land due to the energy received from the sun. The evaporated water rises upwards, condenses in the atmosphere, and forms clouds. The clouds precipitate on the ocean and land in the form of hail, sleet, rain, and snow. Precipitated water on land may go back to the atmosphere by evaporation or improve the moisture content of the soil. A portion of soil moisture may be transferred back to the atmosphere through transpiration, and the remaining water recharges the groundwater by infiltration. Some quantity of water overflows over the earth's surface and reaches the stream channel known as runoff. Snowfall accumulates in mountainous regions and melts in the spring, contributing to streamflow. The water cycle is a complex phenomenon, occurring continuously and simultaneously. The water cycle components can broadly be classified into transportation and storage components which are illustrated in Figure 47. Transporting components are evaporation, precipitation, infiltration, transpiration, and runoff. The storage components are reservoirs, lakes, groundwater storage, soil moisture storage, and Oceans.

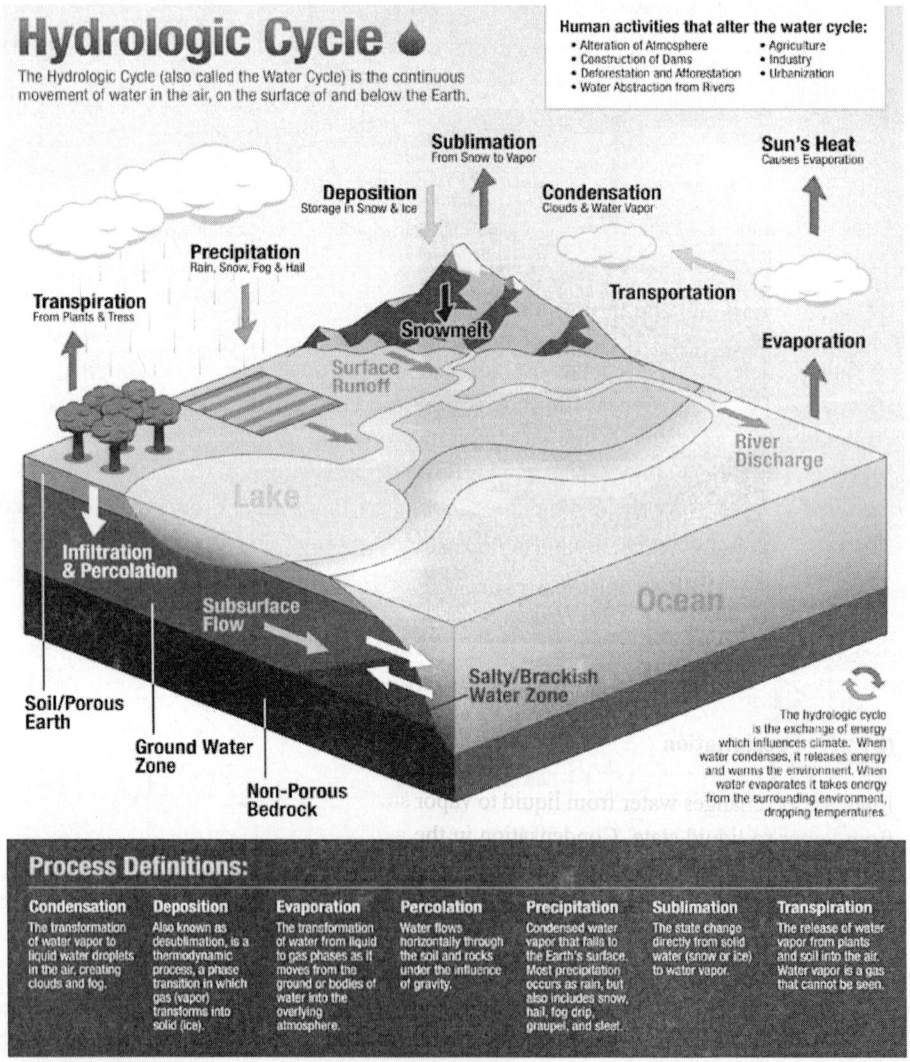

Figure 47: Water Cycle (Source: H2O distributors)

4.2.1. Water cycle components

The driving force of the water cycle is primarily the energy from the sun. The major components of the water cycle are:

(a) Evaporation

Evaporation is the process by which liquid water gets converted to water vapor due to radiation from the Sun. Evaporation is a cooling process during which evaporated water absorbs the heat of the surrounding water, which leads to cooling down. Though many ways exist to measure evaporation, the evaporation pan (Figure 48) is one of the simplest ways with which the daily drop in water level is related to free-water surface evaporation.

Figure 48: Evaporative pan for measuring evaporation (Source: Meted UCAR)

(b) Precipitation

Evaporation changes water from liquid to vapor state, while condensation is the opposite process from vapor to liquid state. Condensation in the atmosphere releases latent energy and converts into water droplets. When the water droplets get heavier, they precipitate on land and the ocean in the form of precipitation. Precipitation is the single most influential factor in the water cycle and a hydrologist must account for the type of precipitation (rain, snow, or hail) when studying a region. Traditionally, precipitation has been measured using rain gauges, which is illustrated in Figure 49.

Figure 49: Rain gauge (Source: Wikipedia)

(c) Infiltration

Infiltration is the downward movement of water through the soil surface. The infiltrated water recharges groundwater supplies and sustains vegetation. The amount of water that enters the soil in a specified time period is the infiltration rate and is influenced by various factors such as soil texture, soil cover, soil temperature, and rainfall intensity.

(d) Interception

Interception is precipitation that is intercepted by leaves and branches of plants and forest floor and doesn't reach the soil. The intercepted water evaporates or infiltrates into the groundwater.

(e) Transpiration

Stored water in the soil is utilized by the plants and vegetation in the photosynthesis process and evaporates from the leaves into the atmosphere is known as transpiration.

(f) Runoff

Runoff is the portion of precipitation that does not infiltrate into the soil. Water that flows at the surface is called surface runoff, and that flows in the soil above the groundwater level is called interflow.

4.2.2. Water Budget Equation

We often need to know how much water is available at a particular location. This can be calculated in a general sense by applying an accounting budget approach to the water cycle components. The volume of water in a watershed can be viewed as the difference between the inflow and outflow and the resulting changes in the storage. Hydrologists call this the continuity equation which is based on the law of conservation of mass.

$$\text{Mass inflow} - \text{Mass outflow} = \text{Change in storage}$$

$$Vi - Vo = \Delta S$$

The water budget equation is simple and universal as it relies on few assumptions of water movement and storage and can be applied to a wide variety of problems at laboratory to continental scales. A basic water budget of a watershed can be expressed as:

$$P - R - G - ET = \Delta S$$

where,

P is Precipitation.

R is Surface runoff.

G is Net groundwater flow out of the catchment.

ET is Evapotranspiration (sum of evaporation from soil, water bodies, and plants).

ΔS is Change in storage.

4.3. Wave and Current system

4.3.1. Waves

The surface of a water body is not static, and the waves also have various origins such as wind, tsunami, and tides. These surface waves together are also known as "gravity waves" as gravity attempts to restore them to equilibrium. Surface waves can be found on the free surfaces of the oceans, rivers, lakes, and ponds. Waves are oscillations generated due to the wind blowing over a water surface, a ship moving on the water, or seismic disturbance at the bottom. Ocean waves are most commonly generated due to wind energy transferring from the air to the water surface.

Based on their driving forces, ocean waves can be classified into the following:

(a) Wind waves

Wind is critical for the marine environment and the response of water to wind (waves) has the greatest impact on ships and coastal structures. Wind waves are created when wind pressure and stress are applied to the water surface. Wind transfers the energy of air molecules to the surface of the water through friction and pressure, which causes the disturbance in the water known as wind waves. The initial stage of wave generation is represented by capillary waves which are waves of small height controlled by the surface tension of water.

(b) Tidal waves

Tide is a generic term for the rise and fall of sea level with respect to the land. Tidal waves are large waves that move over the ocean due to the gravitational influence of the Sun, Moon, and Earth. Tidal waves have a large wavelength. The tidal effect of the Moon is about twice that of the Sun and causes the oceans to swell on either side of the earth, the side nearest to the Moon or the furthest. The tide goes up and down as the earth rotates on its axis.

(c) Tsunamis waves

Tsunamis are long-period waves that are generated due to massive displacement of water due to geological events such as earthquakes, landslides, or volcanic eruptions. The resultant waves travel in all directions from the area of the disturbance. Tsunami waves have extremely long wavelengths, often exceeding 100 kms in the deep ocean. They travel at very high speeds, sometimes exceeding 700 kmph. Tsunamis can have devastating effects on coastal regions due to the amount of water and energy that they carry.

4.3.2. Currents

The oceans of the planet are in constant motion, driven by the force of wind, tides, earth rotation, and changes in the density of water. This vast movement of water is known as currents and they directly impact many human activities on a daily basis such as shipping, fishing, oil and gas activities, etc. The speed of these currents is higher at the top of the surface and decreases with an increase in the ocean's depth. The topography of the bottom of the sea and the shoreline modifies the movement of water, causing currents to change their speed. There are two main types of ocean currents that define the character and flow of ocean waters across the planet: surface and deep-water currents. Although both of them occur by different phenomena, they influence each other.

(a) Surface currents

Surface currents make up only 8% of all ocean water and are generally prevalent in the upper (400 m) of ocean water. The wind and tide drive surface currents in coastal areas, where water rises and falls forcing the water to move. Surface currents in the open ocean are driven mainly due to the wind force. The surface currents move in the same direction as the wind when it blows on the ocean's surface. Because of Earth's rotation, the circulating wind is deflected towards the right in the Northern hemisphere and towards the left in the Southern hemisphere, and this reflection is known as the Coriolis effect. This effect creates circular motion of the surface current that rotates clockwise in the northern hemisphere and anticlockwise in the southern hemisphere, known as gyres (Figure 50). If the Earth does not rotate, the water will move from equator to pole and pole to equator without forming the gyres. As water is good at holding the heat compared to the air, these gyres help in the distribution of warmth. Indian, South Atlantic, North Atlantic, South Pacific, and North Pacific are the five measured gyres worldwide.

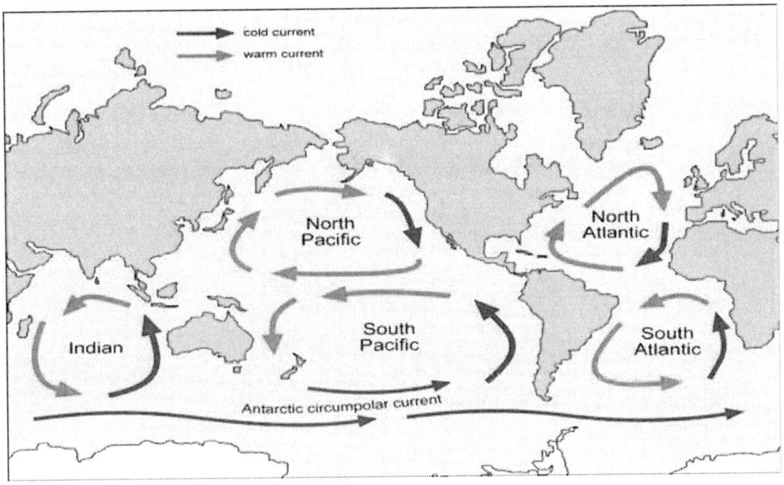

Figure 50: Gyres are circular ocean currents formed by the Earth's wind patterns and the forces created by the rotation of the planet. (Source: Sciencelearn.org.nz)

(b) Deep water currents

Unlike surface currents, deep water currents are driven by the density and temperature gradients. Water becomes colder as it approaches the North pole, forming pure water ice crystals while leaving the salt behind. The water gets heavier due to salt and starts moving downward as the gravitational effect. This water moves along the sea bottom until it rises to the surface. When the deep water moves upward and replaces the surface water, it is known as upwelling, and the reverse is called downwelling. Warm and less dense water comes to the surface, and heavier water goes into the ocean in a process known as thermohaline circulation. Thermohaline circulation and wind-driven surface currents combine and form a loop known as a conveyor belt (Figure 51). Conveyor belt plays a critical role in the Earth's environment. Conveyor belts transfer the oxygen to the deep-water habitats and simultaneously accumulate the nutrients as organic matter and decompose. Climate change can significantly affect the conveyor belt by disturbing the thermohaline circulation. Due to warming climate conditions, polar ice caps are melting, forming a fresh and less saline water surface that will no longer sink.

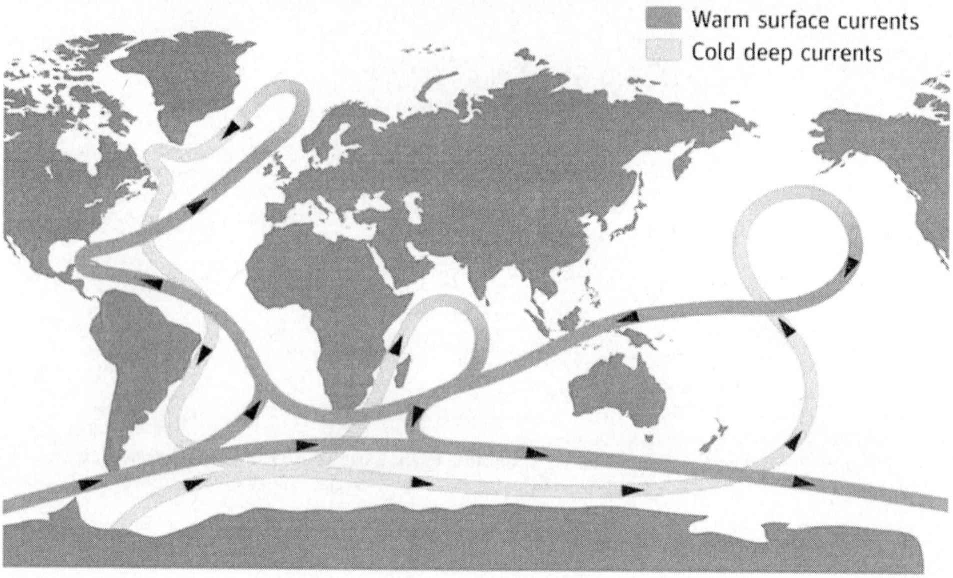

Figure 51: Conveyor belt (Source: exploratorium.edu)

4.4. Sediment transport system

Transportation of solid matters eroded, transported, and deposited by the combined effect of fluid transport and gravitational force is known as sediment transport. Sediment transport is responsible for the transportation of nutrients and contaminants. It contains the eroded minerals from weathered bedrock and organic soil generated from soil formation (Figure 52). Sediment transport occurs naturally in which fluids are water, air, and ice, and transported materials are clay, silt, sand, and grabbles. Sediment transport occurs due to water in the rivers, lakes, and oceans. Erosion, transportation, and deposition are the main three components of sediment transport. Erosion occurs on the Earth's surface and is transported through air, water, ice, and deposited. Based on the transportation mechanism, sediment transport can be classified into four types: aeolian, fluvial, coastal, and glacier.

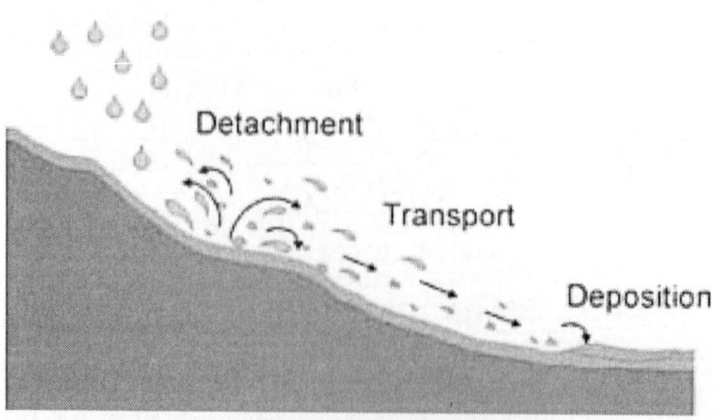

Figure 52: Sediment transport in water (Source: Los Huertos, M. (Ed.), 2020.)

Solid particles transported due to wind forces are called Aeolian sediment transport. Smoller solid particles are transported in aeolian transportation as air density is very low, and transported solid particles form bedforms, ripples, and dunes. Aeolian sediment transport most occurs in the arid regions.

4.5. Water Resources Structures

Water resources engineering is the quantitative study of the hydrologic cycle. Its applications include the design of hydraulic structures such as dams, spillways, sewage conduits, waterways, erosion protection structures. In general, a water resource structure is a building that is completely or partly submerged in water and changes the way water flows naturally. The flow is diverted, slowed down, controlled, or stopped completely by these structures. A hydraulic structure can be built in a river or any other body of water where the natural flow of water needs to be changed. Below are a few examples of water resources structures and the functions they serve.

4.5.1. Dams and Reservoirs

Reservoirs are used for storing water and they can serve various purposes such as conservation, flood mitigation, maintaining year-round water supply, hydropower generation, etc. If the reservoir was built for one purpose only, it is called a single-purpose reservoir, and if it is built for multiple purposes, it is called a multi-purpose reservoir.

A multi-purpose reservoir could be built to protect communities downstream from flooding and also save water for agriculture, water supply, hydroelectricity, and other uses. For example, the Bhakra Nangal dam in Punjab, one of the highest concrete gravity dams in the world, forms the Gobind Sagar reservoir and is intended for irrigation and hydroelectric purposes. It was famously described by India's first Prime Minister Jawaharlal Nehru as "New Temple of Resurgent India" highlighting the importance of large hydraulic structures in the economy.

Most dams in India are funded and maintained by the government along with a few other organizations such as National Hydro Power Corporation (NHPC), Damodar Valley Corporation (DVC), etc. Following are the most common benefits of a multi-purpose reservoir depending on the nature of the project.

(a) **Hydropower generation:** Energy is important for the social and economic success of a country and hydropower is a cost-effective renewable energy source. Hydropower is electricity made from water stored in reservoirs using turbines. Depending on how much electricity can be generated, it can be used to meet regional or national needs.

(b) **Flood control:** River floods cause immense damage to life and property every year. Projects with explicit mandate for flood control can impound floodwaters and release them under control to the river downstream of the dam or store or divert the floodwater for other uses.

(c) **Irrigation:** Reservoirs store extra water during the rainy season which can be used to irrigate dry agriculture areas. Using canals, the flow of water can be regulated to meet the farming needs of different areas at different times of the year. Irrigation is one of the primary benefits of multi-purpose projects.

(d) **Drinking and Industrial Use:** Reservoirs can play a crucial role in ensuring stable water supplies, especially in areas that experience droughts in certain parts of the year. Essentially, reservoirs are intended to store water during the high flow season and use it during the low flow season based on demand. The stored water can also be used to meet various industrial demands.

(e) **Water inland navigation:** Such projects can also enhance inland navigation. Transport of heavy goods via main rivers or canals is a cheaper mode of transport than roads. Comprehensive planning and development of such projects can yield major economic benefits.

(f) **Recreation and Fishery:** Reservoirs also have many benefits such as promoting fishing and tourism. During the design process, the leisure benefits of reservoirs such as boating, swimming, fishing, etc. should be taken into account so that the project accrues economic benefits to the local community and businesses.

Along with its benefits, multi-purpose reservoir projects also come with significant social and economic costs related to human resettlement and damage to the environment and forests. Reservoirs often drain or submerge large swathes of cultivable or forest land. If land is flooded, the communities settled there would have to be compensated or resettled elsewhere which would adversely affect the local economy and culture. For the overall success of such projects, proper studies should be conducted on post-construction socioeconomic and environmental impact assessment before embarking on any construction and adequate measures implemented to minimize the damage.

4.5.2. Marine Structures

The utilization of the oceans as a means of transportation is crucial to international trade. There are huge reserves of raw materials such as hydrocarbons in the waters and on the ocean floor. The seas hold a lot of promise for exploration as well as for fish and plankton. Seventy percent of the sun's radiation reaching earth is absorbed by the seas. Thermal gradients, wind, current or wave energy, salt gradients, and other forms of this energy can all be captured. However, it takes marine structures to take advantage of these opportunities. Marine structures are large, expensive, and complex structures located in a hostile environment, necessitating study and qualified engineers for their design and construction. Research and engineering are thus crucial for any nation desiring to take part in industrial progress in this field. Often termed as maritime structures, marine structures encompass various infrastructure that is built in the ocean so that marine resources can be utilized.

Based on how they work and what they look like, marine structures can broadly be classified as coastal, offshore, and deep-sea exploration structures. Maritime structures differ from normal structures because:

(a) Testing of prototypes is not usually possible and therefore first principle-based analysis and designing needs to be done.

(b) Only probabilistic expressions of load and the response is possible.

(c) Very high cost.

(d) Considered as global activities and thus international laws and regulations are applicable.

(e) The ocean or the sea has some serious challenges in terms of construction material.

Coastal structures are often built to stop the deterioration of landscapes and infrastructural facilities along the coastline and to minimize the risk to people and businesses that depend on the coastal area. They are generally made of materials such as concrete, large armor stones, steel, or wood. However, use of high strength concrete, pre-stressed concrete, concrete that does not freeze, and concrete that is light in weight is also increasing.

Oil and gas will continue to be a major source of energy for the foreseeable future. To exploit such sources located much below the sea level, marine offshore structures are required. The design and planning of such structures is usually done for a specific location due to very high cost and detailed designing. Due to the high risk of fires and explosions seen historically, design of offshore structures should consider additional factors of safety. Levees, seawalls, etc. are the most common examples of coastal structures. Offshore structures are highly susceptible to corrosion and thus the construction of such structures requires significant technical expertise.

The Mumbai High Field is an example of such a structure (Figure 53). It is an offshore oilfield 176 kms off the west coast of Mumbai and its operations are run by the Oil and Natural Gas Corporation (ONGC).

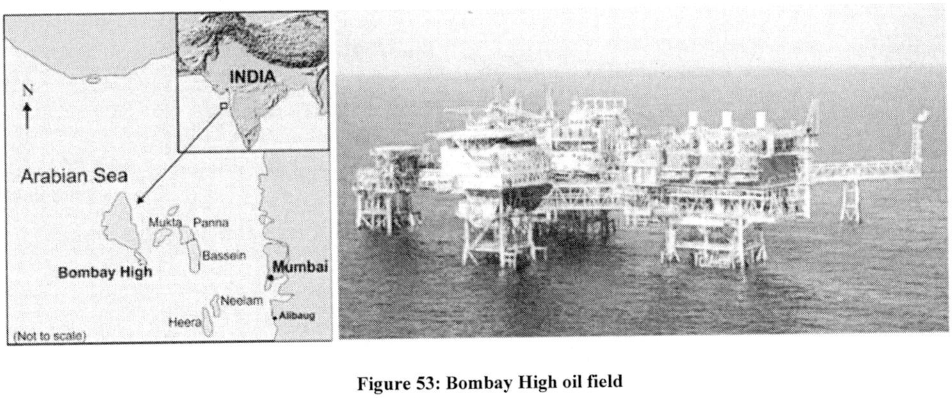

Figure 53: Bombay High oil field

4.6. Water Treatment and Supply

People impact the environment in myriad ways. One of the most common ways human activities impact the environment is by polluting water. According to the World Health Organization (WHO), between 50 to 100 liters of water is required per person per day. Thus, water resources engineers have a major role to play in designing and operating treatment and supply systems in order to meet the demands of the population.

4.6.1. Water Treatment Systems

The main goal of water treatment and purification is to access natural water from the best possible sources and put it through a physio-chemical process that ensures good physical quality, no bad taste or smell, and causes no harm to public health. Water can be obtained by drawing directly from rivers, lakes, or springs. Depending on the source, they vary greatly in purity and suitability. Underground water may contain a significant amount of matter dissolved from the state through which it passes. Shallow wells are much more susceptible to bacterial contamination than deep wells. Streams and rivers carry impurities from the earth and from the discharge of sewage and industrial effluents to ponds, lakes, and reservoirs. There have been numerous cases of recurring epidemics traceable to drinking water supplies in the past.

A good water treatment system deals with the physical and chemical characteristics of water. The following are the Physical and Chemical characteristic of water.

Physical characteristics	Chemical characteristics
• Turbidity • Color and Temperature • Taste and Odor	• Total solids • Alkalinity • PH Value • Hardness of Water • Chloride content • fluoride content • Metal and other chemical substances • Nitrogen and its compounds • Dissolved gasses

Each of these undesirable physical and chemical characteristics can be treated up to different limits. A few basic processes that are often followed by water treatment systems are as follows (Figure 54):

(a) **Screening:** Generally screening is done in order to eliminate particles such as debris, trees, animals, etc. Coarse and Fine screens are used based on the size of particles.

(b) **Plain Sedimentation:** If the water contains large-sized suspended impurities, preliminary sedimentation is a very cost-effective way to remove them. Because suspended impurities cause turbidity in the water, removing them results in more uniform water that can be used in subsequent treatment processes. Plain Sedimentation is the process of getting particles that are floating in water to settle to the bottom by keeping the water still in tanks.

(c) **Sedimentation with coagulation:** By adding coagulants to the water flow, very fine colloidal particles will be produced. The very fine colloidal particles in water are attracted and absorbed by these flocs, resulting in larger flocculated particles. Coagulants such as alum, copperas, chlorinated copperas, and sodium aluminates can be added to sedimentation tanks in water treatment.

(d) **Filtration:** The water that is filtered through sediment may contain bacteria and very small particles. The water is filtered through beds of granular material, like sand, to get rid of or reduce any remaining impurities. The process of filtering is to let water flow through beds of such granular materials.

(e) **Disinfection:** In order to disinfect water of bacteria, the most commonly used method is boiling and chlorination. There are many more ways to accomplish this by treatment with lime, UV rays, potassium permanganate, iodine and bromine, the choice of which is decided based on cost effectiveness and operational requirements.

(f) **Aeration:** Aeration treatment is when a lot of air is pumped through water and then let out of the water. The air makes the gasses or compounds that are dissolved in the water rise to the surface which are then let out. For example, high concentrations of iron and manganese can be treated with an aeration and filtration system. After the air is mixed with the passing stream of water, water enters an aerator vessel where air is separated from the water. This water then flows through a filter where various filter media screen out oxidized particles of iron and manganese.

(g) **Softening:** Softening is the process of removing hardness from water. Hardness is a water property that prevents soap from lathering. There are two kinds of hardness. Temporary hardness is because of the presence of sulfates and carbonates of calcium and magnesium which can be removed by boiling. Permanent hardness is caused by nitrates, calcium chlorides, and magnesium chlorides in the water and may be removed using the zeolite method.

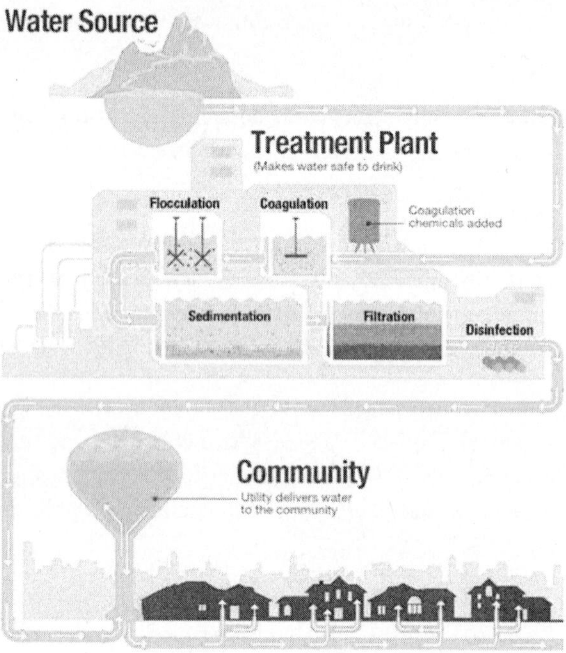

Figure 54: Water treatment plant for domestic supply (Source: cdc.gov/drinkingwater)

4.6.2. Effluent treatment systems

Effluent, also known as trade effluent or wastewater, is treated sewage from a septic tank or a sewage treatment plant. It is any kind of waste that does not come from kitchens, surface water, or homes. Industrial and commercial buildings are major producers. Most wastewater flows directly from a building into the main sewer system and cannot be allowed to enter a river, reservoir, stream, or lake until it has been cleaned and treated. An effluent treatment plant is where industrial wastewater is treated based on prevalent environmental standards. And effluent treatment is crucial to ensure sustainable economic development of industrial areas. The various stages of an industrial effluent treatment plant are shown in the Figure 55.

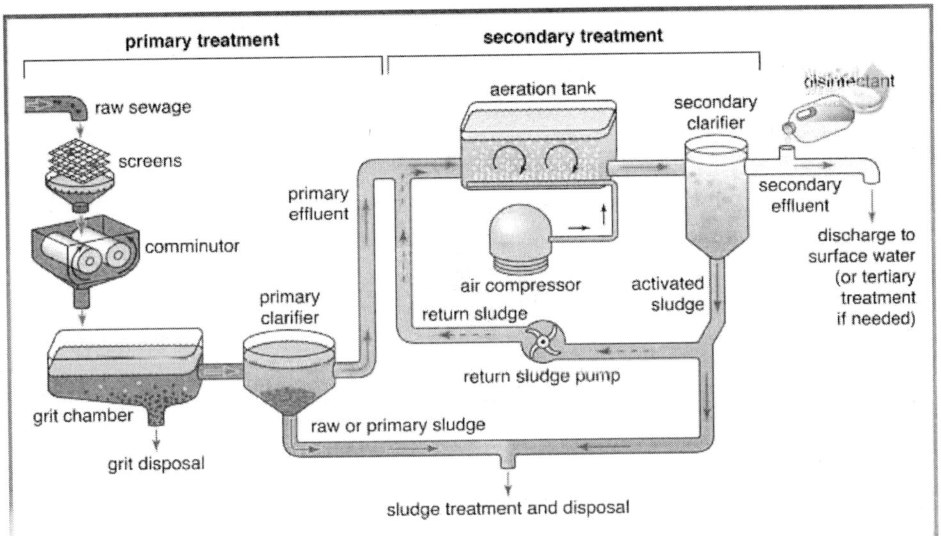

Figure 55: Stages of Industrial Effluent Treatment Plant (Source: Netsolwater)

There are four levels of treatment in any effluent treatment system.

(a) **Preliminary Treatment:** Large impurities such as cloth, plastic, wood logs, paper, etc. are physically separated. The common unit operations at the preliminary level are screening, sedimentation, and clarification. In order to get rid of large solids such as plastic and clothes, we utilize a screen of maximum 10 mm size. Further, sedimentation involves using gravity to remove the solids that are suspended in the water. To make the water clearer, clarification is performed that separates solids from fluids.

(b) **Primary Treatment:** Materials that float and sink, like solids that are suspended and organic matter, need to be taken out. At the primary level both physical and chemical treatments are done. Chemical coagulation and flocculation, pH treatment is some of the most common treatments in this segment. Chemical processes are aided always by physical operations and sometimes biological treatment can also be performed. Addition of chemicals to bring out changes in the quality of wastewater is the basic idea behind the primary treatment. Coagulants such as alum, iron sulfates are used along with some flocculants to ensure the primary treatment.

(c) **Secondary Treatment:** In this part of treatment, biological processes around microorganisms, specifically bacteria, are involved. This is generally done in order to reduce and remove mainly organic and some inorganic compounds that might be present. There are two types of treatment processes in secondary treatment namely:

- Aerobic: Treatment process taking place in abundance of oxygen i.e., presence of air. Such processes involve microorganisms that use free air (oxygen) to break down organic impurities into carbon dioxide, biomass, and water.

•Anaerobic: Such treatment processes take place in the absence of air. They employ microorganisms that do not require free air (oxygen) to break down organic impurities.

(d) Tertiary Treatment: The purpose of advanced or tertiary treatment is to improve the waste-water quality before it is finally reused by removing any remaining substances or inorganic compounds like phosphorus, nitrogen etc. At this step, microbes, pathogens, and parasites that are bad for population health are also removed.

4.6.3. Water Supply Systems

A functioning water supply treats and removes impurities to make it safe for public consumption. The goals of a community water supply system goals are as follows:

•To provide consumers with a sufficient amount of water for drinking.

•To ensure people have enough to meet their basic needs along with contingencies such as firefighting, festivals, meetings, etc.

•To plan for future demands due to population growth, rising living standards, storage, and transportation.

•Preventing water pollution at the source, making the storage clean and hygienic, and ensuring a safe and clean transport system for water.

•Ensure the treatment units and distribution network are efficient, cost-effective, and dependable.

A water supply system includes all facilities for water transport from the source, treatment, and then to the distribution. A domestic water-supply system is a way for public utilities to get water of the right quality, quantity, and pressure from a source to a home. This is usually done with a system of pumps and pipes. The four major types of water distribution systems are as follows:

(a) **Tree or dead-end distribution system:**

In this type of water distribution system, a main central line runs through the buildings. From this main line, various sub-mains branch off on either side. Further, these sub-main lines get divided into various branch lines from which service connections are provided for the households (Figure 56).

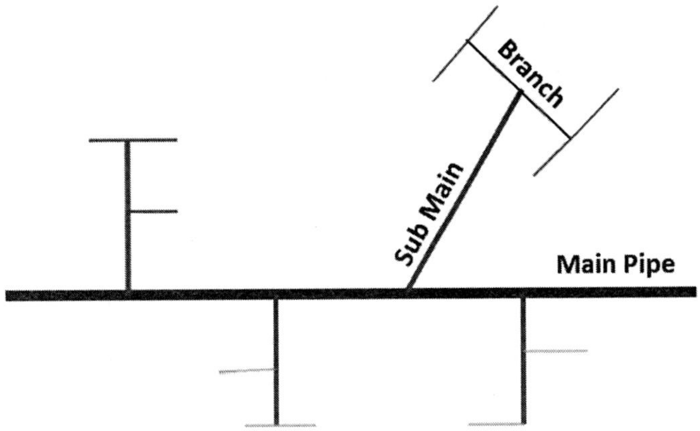

Figure 56: Dead end water distribution system (Source: SWMM)

(b) Grid Iron Distribution System:

Main supply lines in this type of distribution system run along the centerline of the structure. The sub mains branch off in perpendicular directions. The main branch acts as a link between all the sub-mains. The unique characteristic of such a system is that all the pipes are interconnected (Figure 57). Dead ends do not exist in such a system. Water can come in from different directions at the point of pullout, which makes it easier to work, especially when repairs are needed.

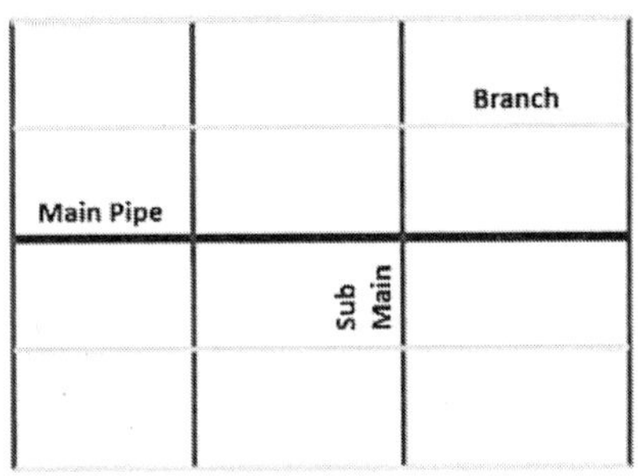

Figure 57: Grid iron water distribution system (Source: SSWM)

(c) Circular Distribution System

The main pipeline is set up around the city or area in a ring system, which can also be called a circular system. The branch lines go in the opposite direction from this main line, and they also connect to each other. So, every street in the area will get just the right quantity of water. Circular system is better for a town with roads and streets that are well laid out. The circular water distribution system is illustrated in Figure 58.

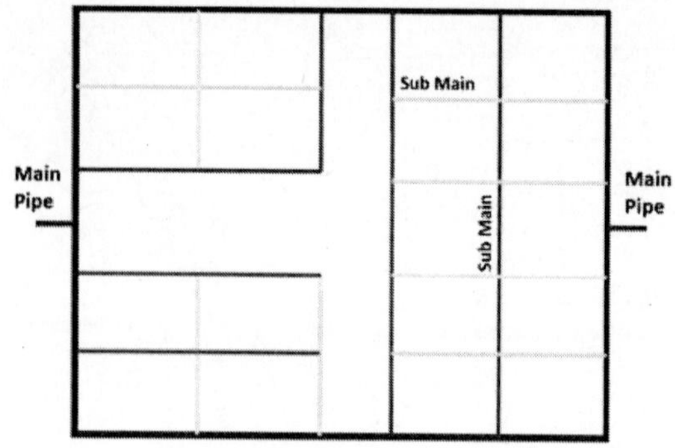

Figure 58: Circular water distribution system (Source: SWMM)

(d) Radial Distribution System

The entire building is divided into several distribution areas in this type of water distribution system. There is an elevated reservoir that is centrally located. From this reservoir, distribution pipes run radially towards the areas in the distribution periphery. Such a system provides prompt service with minimal disruption (Figure 59). The system of laying pipes is very simple in the radial design distribution system.

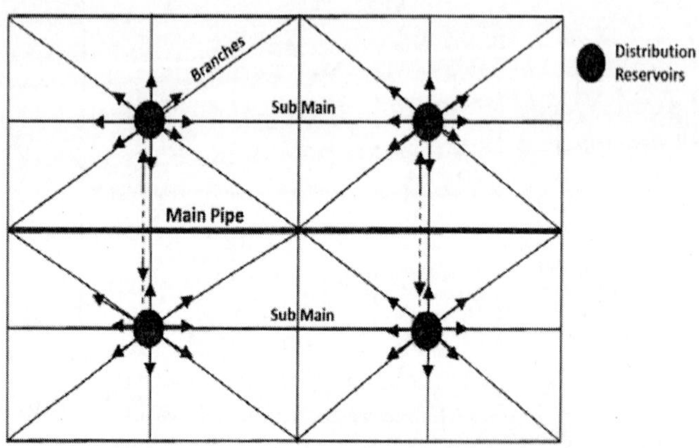

Figure 59: Radial water distribution system (Source: SWMM)

Water resources engineering is a vast and dynamic field where our rapid and significant advances in our understanding of physical processes leads to improvements in design and implementation of various engineering structures.

References and suggested readings

1. *Eeltink, D., Armaroli, A., Brunetti, M., Kasparian, J., 2020. Reconciling different formulations of viscous water waves and their mass conservation. Wave Motion 97, 102610. https://doi.org/10.1016/j.wavemoti.2020.102610*
2. Bansal, D. R. (2010). *A Textbook of fluid mechanics and hydraulic machines.* New Delhi: Laxmi Publication (P) Ltd.
3. Los Huertos, M. (Ed.), 2020. Ecology and Management of Inland Waters. Elsevier, pp. i–iii. https://doi.org/10.1016/B978-0-12-814266-0.00002-7
4. Dr. P N Modi, D. S. (2017). *Hydraulics & Fluid Mechanics including Hydraulics Machines.* New Delhi: Standard Book House.
5. Ranald V Giles, J. B. (1994). *Theory and problems of Fluid Mechanics and Hydraulics.* Schaum's Outline Series, McGraw-Hill, Inc.
6. Joseph A Schetz and Allen E Fuhs. Fundamentals of fluid mechanics. John Wiley & Sons, 1999

Dynamic QR Code for Further Reading

1. Simulations of fluid flow: https://physics.weber.edu/schroeder/fluids	Fluid Flow
2. Water cycle and other precipitation related materials by NASA: https://gpm.nasa.gov/education/water-cycle	Water Cycle

3. Dam construction video. https://youtu.be/OUpZHFx-GxM	Dam
4. Wastewater treatment video: https://youtu.be/bDW6XbQhTKI	Wastewater
5. Water distribution video: https://youtu.be/yZwfcMSDBHs	Distribution

5 Traffic and Transportation Engineering

Unit specifics

Through this unit we have discussed the following aspects

- *Introduction to transportation: Different modes and integration*
- *Major transport infrastructure developments in India*
- *Introduction to geometric design*
- *Public transport systems*
- *Transportation logistics*
- *Road safety*
- *Advanced technologies and ITS/IT/IoT*
- *Public-Private Partnership in transportation.*

The practical applications of the topics are discussed for generating further curiosity and creativity as well as improving problem solving capacity.

After the related practical, based on the content, there is a "Know More" section. This section has been carefully designed so that the supplementary information provided in this part becomes beneficial for the users of the book. This section mainly highlights the initial activity, examples of some interesting facts, analogy, history of the development of the subject focusing the salient observations and finding, timelines starting from the development of the concerned topics up to the recent time, applications of the subject matter for our day-to-day real life or/and industrial applications on variety of aspects, case study related to environmental, sustainability, social and ethical issues whichever applicable, and finally inquisitiveness and curiosity topics of the unit.

Rationale

This fundamental unit on transportation engineering introduces basic concepts related to various aspects of transportation engineering. This chapter begins by explaining different modes of transportation and their role in India's development. It then provides an introductory guide to various components of transportation engineering like geometric design, pavement materials and construction, public transport systems, transportation logistics, and road safety. The last section of this chapter discusses different types of Public-Private Partnership adopted in the transportation sector for developing the road infrastructure. This chapter introduces different concepts of transportation engineering to give a basic idea of the subject and to explain the role of advanced technologies in the development of the transportation sector.

Transportation engineering deals with the road design, construction, operations, and maintenance. It also deals with the management of traffic, planning and operation aspects of public and private transport systems. Transportation engineering is the branch of civil engineering that studies the different methods of road design and construction, traffic management, public transport systems, and road safety using scientific and engineering techniques.

Pre-requisites

Basic elements of civil engineering

Mathematics (Class X)

Unit outcomes

List of outcomes of this unit is as follows:

U1-O1: Understand basic concepts of different modes of transportation.

U1-O2: Understand the major development in the transportation sector in India.

U1-O3: Understand the basic concepts and components of geometric design of roadways.

U1-O4: Understand the basic elements of pavement design and construction.

U1-O5: Understand the basic concepts of public transport systems.

U1-O6: Understand the basic ideas of transportation logistics.

U1-O7: Introductory guide to road safety aspects.

U1-O8: Understand the ideas and components of intelligent transportation systems.

U1-O9: Understand the role of Public-Private Partnership (PPP) in transportation and its role in the development of transportation infrastructure.

5.1. Introduction to transportation: Different modes and integration

Transportation is the movement of people and commodities from one location to another. Transportation can be done through (a) land, (b) air, and (c) waterways. Ancient transportation was mainly through land using human-powered (walking, palanquin, and pulled rickshaws) and animal powered (e.g., bullock carts and tangas) transports. Land transport was developed to accommodate vehicles on road and locomotives on rail. Humans explored water as the possible mode of transport to mainly transfer commodities from one place to another place. Humans also developed airways for quick transport of people and commodities. At present, roadways, railways, waterways, and airways are the major modes of transportation.

Pipelines, aerial ropeways, elevators, cable cars, belt conveyors, and hyperloops are the other possible modes of transportation. Pipelines are extensively used for transporting liquids (water, crude oil, etc.), gas (natural gas, steam, etc.), and in rare cases solid (solids converted into slurry) particles as well. Belt conveyors, cable cars, and aerial ropeways are used for transportation at some altitude for a short distance. Hyperloop is the latest and one of the quickest modes of transport for humans. The following sub-sections briefly explain the four major modes of transportation.

5.1.1. Road transportation

Land transportation is the most common mode of transportation. Road system is systematically used by pedestrians and various classes of vehicles like motorized two-wheelers, cars, trucks, buses, etc. Road transportation provides maximum flexibility to the traveler in terms of route choice, time, and speed of travel. Over the years, road infrastructure has been substantially developed for providing end-to-end connectivity. Thus, road transportation is the most convenient way for door-to-door delivery. The remaining modes of transportation have to depend on road transportation at some point in order to complete the delivery service or to reach the desired destination. Therefore, road network must be systematically planned and developed throughout the country so that it can operate independently as well as act as a feeder system for the remaining transportation modes.

The main downside of road transport is the weather, traffic, and rising cost of petroleum energy. The rate of road transport of goods gets adversely affected during the rainy season as compared to the summer season. Further, increase in traffic on roads significantly raises the rate of emission pollutants than other transportation modes. Additionally, the rising cost of petroleum energy is rapidly increasing the cost of road transportation. Therefore, it is important to adopt green energy for reducing air pollution as well as for reducing the dependency on petroleum energy.

5.1.2. Rail transportation

Rail transportation was first invented in the early 19th century and since then has played a crucial role in transporting passengers and bulk commodities over long distances for the past two centuries. Rail transport consists of a locomotive (engine) with a series of passenger bogies and/or wagons moving on a fixed path defined using two parallel steel rails of the railway track. In early days, rail transport used to work on steam locomotives where water was converted to steam by burning combustible materials like wood and coal. Later on, steam locomotives were replaced by internal combustion engines which were powered by diesel (a petroleum energy product). In recent years, most of the locomotives are powered using electricity by providing an overhead line or by providing on-board energy storage using batteries. Locomotives using steam or diesel as a power source generate tremendous air pollution. This was one of the main reasons behind India adopting electric locomotives for rail transportation.

Rail transportation is a comfortable and economical mode of choice for transporting passengers as well as commodities over a long distance. Railways serve as a mass transit system for the commuters traveling inter-city as well as intra-city. However, rail transport gets confined to a limited path providing minimal chance of divergence. The passengers and transfer of commodities to the end point have to rely on road transport. Rail transport is relatively faster than road transport because railways are not influenced by traffic, point of diversion, and switch off between the nodes. These attributes make rail transport more reliable than road transport for making long hauls with nominal damage.

Indian railways are considered as a lifeline of the nation providing comfortable and cheap transport across the country. Indian railways have one of the largest rail networks in the world with 67,956 kms of route length. 94.10% of the total route kms consists of broad gauge, 3.54% and 2.36% of the total route kms consists of meter gauge and narrow gauge, respectively (Indian Railways Civil Engineering Portal, 2022). Indian railways carried 8086 million passengers and 1208.41 million tonnes of freight in 2019-20 year. Indian railways provide inter-city and inter-state travel (Figure 60) whereas various government and semi-government agencies operate metro (Figure 61) and monorail (Figure 62) services for intra-city travel for the passengers. Interested readers are requested to refer Indian Railways Civil Engineering Portal (2022) for exploring the major rail infrastructure projects currently undergoing or recently completed projects.

Figure 60: Indian railways (Indian Railways Civil Engineering Portal, 2014)

Figure 61: Delhi metro (Delhi Metro Rail Corporation, 2022)

Figure 62: Mumbai monorail (MMRDA, 2013)

5.1.3. Water transportation

Water transportation is the main transport mode for global trading. Almost 90% of the global trade is conducted through waterways. This is mainly because water transport offers minimum resistance to traction which results in less energy consumption to haul unit load through unit distance. Therefore, water transport serves as the cheapest mode of transport for trading the highest volume of freight. However, water transport is the slowest among the major four modes of transport. Because of this reason, waterways are mainly used for freight transportation. Water transport is mainly used for transporting chemical, hazardous and toxic products, machinery and factory parts, vehicles and their accessories, petroleum products, livestock and animals, and food.

Harbors and ports are constructed to operate container ships for parking, loading-unloading, and storage of cargo. Harbor is a section along the coastline where ships and containers or other water vessels are parked and stored. Port is a commercial water facility built around the harbor for loading and unloading cargo or other water vessels. Transport planning is important while developing harbor and port. All the freight to be transported through waterways will be initially transported using road or rail transport. Thus, road and rail transport serve as a feeder system for transferring passengers and commodities back and forth the harbor and port.

Water transport can also be operated within a country by developing inland water transport. The Government of India established Inland Waterways Authority of India (IWAI) on 27 October 1986 for developing inland waterways for shipping and navigation. India has developed 20,275 kms of national waterways across 24 states in the country comprising rivers, canals, creeks, backwaters, etc. India efficiently transfers around 55 million tonnes of cargo every year through inland waterways (Ministry of Ports Shipping and Waterways Government of India, 2018). Figure 63 shows the water transport operated at the coastal regions of India.

Figure 63: Indian Waterways (Ministry of Ports Shipping and Waterways Government of India, 2018)

5.1.4. Air transportation

Air transportation is an integral part of the services sector for achieving economic growth of the country. Air transport plays a pivotal role in the integration of local economies with the global economy by providing critical connectivity on a regional, national, and international level. Air transport is the quickest among the four major transportation modes. Air transport provides uninterrupted passage over land and water which has reduced distances by decreasing travel time. Air transport is the most convenient and comfortable mode of transport for passengers aiming to reach their destination as quickly as possible and save substantial time which might be lost while traveling through other modes. Air transport also plays a key role in commodities transfer. High-value shipment on long hauls in less time is possible through air transport. The major limitation of air transport is high fuel cost and high overall operating costs making it the most expensive mode of transport among the four major modes. Another limitation is that air transport cannot be operated during extreme weather conditions like rainstorms and thick fog.

Airports facilitate the use of airplanes for transporting passengers and commodities. Airport consists of a runway, taxiway, terminal facilities, air traffic control centers, hangers, visual aid, and safety systems. Road and rail transport can serve as a feeder system for transferring passengers and commodities back and forth the airport.

5.1.5. Intermodal transportation

Intermodal transport is an integral part of freight transportation. Land (road and rail) transport serves as a feeder system for water and air transport. Intermodal freight transport involves an intermodal container or a vessel which will be transferred from one location to another using multiple modes of transportation until it reaches the final location. For example, India exports gems and jewelry to countries like the USA, Hong Kong, UAE, Belgium, and Israel to name a few. The shipment of gems and jewelry in India will most likely begin on road, then to air or water transport, which may transfer to rail in foreign countries such as the USA and then it arrives at its destination by road. The purpose of intermodal freight transport is to optimize the cost of transport by taking advantage of the individual strengths of the four major modes of transport. Figure 64 shows a typical representation of intermodal transportation.

Intermodal transport also plays a key role in passenger transport over long distances. The passenger selects the mode of transport based on comfort, cost of transport, convenience, and travel time. A passenger having no time constraint may choose a train or a ship journey based on his/her comfort, convenience, and cost of travel. A passenger having time constraints may overlook the cost of travel and choose air transport to reach the desired location on time. Here, it can be noted that road transport will most likely serve as a feeder system when the passengers need to travel long distances.

Figure 64: Intermodal transportation (DHL Logistics of Things, 2022)

5.2. Major transport infrastructure developments in India

India's transportation sector is huge, catering more than 1 billion people on a daily basis. Transportation is one of the major sectors contributing to India's GDP (Gross Domestic Product) growth. The services sector in the year of 2019-20 was expected to grow at 6.9% with a contribution of 55.3% to India's gross value added (Saket, 2020). Therefore, it is important to invest and develop India's transport infrastructure to support and manage the ever-rising demand of passenger and freight transportation within as well as outside the country.

5.2.1. Road transport infrastructure development

The Ministry of Road Transport and Highways (MORTH) handles development, challenges, maintenance, and issues occurring in road transport infrastructure in India. In recent years, MORTH has started focusing on improving the NHs (National Highways) network by developing road connectivity in border areas, coastal roads, major and non-major ports. The MORTH is also focused on developing proficiency of national, inter, economic, and feeder corridors through Bharatmala Pariyojana programme (Ministry of Road Transport and Highways, 2017). The Government of India has proposed an investment of INR 5,35,000 crores in Phase I over 5 years. Bharatmala Pariyojana programme envisions development of a 24,800 km road network of economic corridors along with GQ (Golden Quadrilateral) and NS-EW (North-South and East-West) corridors for expediting freight transportation via roads. The Phase I of Bharatmala Pariyojana programme will develop a length of 9,000 km for economic corridors, 6,000 km of inter-corridors, 5,000 km of national corridors, 2,000 km of border roads, 2,000 km of coastal and port roads, and 800 km of expressways. The programme will also develop 10,000 km of remaining road works under NHDP (National Highways Development Project). The programme will also try to counter traffic congestion problems commonly faced in urban cities by constructing ring roads and elevated (bypass) corridors along with lane widening projects.

5.2.2. Rail transport infrastructure development

As mentioned earlier, Indian railways is considered as the lifeline of the nation. Bulk freight transport and passengers required to travel across places prefer rail transport as a feasible option. Indian railways had a lot of challenges like over utilization of the infrastructure (more than 60% of the routes were utilized over the capacity), scarce freight carrying capacity, organization stringency, and insufficient focus on customers (Indian Railways, 2017). Considering all these issues, the Indian Government developed a five-year plan for reducing rail network congestion and expansion of rail network, improving North-eastern and Kashmir connectivity, enhancing safety via track renewal and construction of bridge works, introducing high-speed rail and elevated corridors, and improving logistics parks. Indian railways increased broad gauge lines from 1,983 km to 3,000 km in the span of two years. The Government of India invested around INR 24,000 crores for developing dedicated freight corridors. Nearly 2,000 km of dedicated freight rail network was completely electrified in the span of two years. The Indian Government introduced one of the fastest trains 'Gatiman express' for swift transport. Talgo trains will also be included in Indian railways which will operate on the existing infrastructure for fast transportation. Indian railways are electrifying the rail network at almost 2,000 kms/year for having a pollution free, energy efficient, and environment-friendly mode of transport. Indian railways are also focusing on operating semi-high-speed trains along GQ. Indian railways are determined to develop infrastructure using innovative technologies and digital services for delivering safe, efficient, and high-performance rail transport systems for the customers.

5.2.3. Water transport infrastructure development

The Ministry of Ports, Shipping, and Waterways has decided to accelerate the growth of the maritime sector in the coming decade. A project named 'Maritime India Vision 2030 (MIV 2030)' is introduced to strengthen the coastline by giving top priority to port-led development (Ministry of Ports Shipping and Waterways, 2021). India has a 7,500 km long coastline of navigable waterways with 12 major and more than 200 non-major ports. Maritime sector handles around 95% of the country's trade volume and 65% of the trade value. Jawaharlal Nehru Port Trust (JNPT) and Mundra port are the two Indian ports featured in the list of top 40 global container ports. In the last 5 years, India increased its capacity of handling cargo by 65% across all major ports. India started an ambitious and flagship programme named Sagarmala to revolutionize India's logistic sector performance by making full use of India's coastline and waterways. The Ministry of Ports, Shipping, and Waterways has identified more than 574 projects for port modernization and new port development, port connectivity enhancement, port-linked industrialization, and coastal community development during 2015-2035. A total of 236 ports are selected for modernization from which 68 projects are completed and 70 are under implementation. 235 ports' connectivity will be developed through the Sagarmala project. Currently, 35 projects are completed, and 94 projects' work is under implementation.

5.2.4. Air transport infrastructure development

The Ministry of Civil Aviation, Airports Authority of India handles development, challenges, maintenance, and issues occurring in air transport infrastructure in India. India has over 450 airports, out of which 125 airports are government-owned Airports Authority of India (AAI) (Ministry of Civil Aviation, 2019). AAI covers around 9.6 million sq. km and is liable for providing Air Navigation Services (ANS) in Indian airspace and in the Indian ocean region. In total, 15 airports in India manage around 83% of the passenger throughput in the country. Increasing airport infrastructure capacity is one of the main objectives of the country. Delhi airport has three terminals, and the country is planning to add an additional runway in 3-4 years. An additional airport will be developed in Mumbai, Pune, Vizag, and Goa to handle ever-rising growth in passengers. Indian airways also made efforts to reduce dwell time for air cargo from 72 hours to 48 hours. In fact, nearly 25% of the total dwell time has been reduced at seven airports handling import and export cargo in the last 3 years. There exists further scope to decrease the dwell time by adopting paperless processing, infrastructure improvements at cargo terminals, and use of off-airport cargo processing. It has been projected that nearly 821 million domestic passengers will travel within India and around 303 million international passengers will visit India and air cargo throughput may rise from 3.35 mmtpa (millions metric tonnes per annum) to 17 mmpta by FY 2040. Therefore, it is important to develop the airport infrastructure of the country. Considering this aspect, the Government of India has launched UDAN Yojana to ensure appropriate infrastructure development, affordable air travel, and implement regional connectivity for ensuring economic growth of the country.

5.3. Introduction to geometric design

Geometric design is one of the important topics in Transportation engineering, dealing with dimensions and layout of the roadway. The prominence of geometric design is to provide maximum safety and cost- efficiency in traffic operations (Kadiyali, 2013). The key features of geometric design are cross-sectional elements, provision of sight distance, horizontal and vertical alignments. The geometric design is mainly influenced by various factors such as design speed, topography, vehicle characteristics, traffic flow, environmental factors, and economy. The crucial factors are discussed in the following subsections.

Design speed

The most crucial component affecting geometric design is design speed. Design speed directly influences sight distance, horizontal and vertical curve designs. India has more than 20 vehicle categories which can be driven at different speeds as per the driver's perception and comfort. Therefore, it is very difficult to select a particular speed value for designing geometric features as per type of vehicle or driving behavior. Therefore, an 85th percentile speed is normally considered for designing road geometric features. However, in some countries, 98th or 95th percentile design speed is also considered.

Other factors

• Topography: Topography of the terrain can be categorized into four parts: plain (0-10% cross-slope), rolling (10-25% cross-slope), mountainous (25-60% cross-slope), and steep (60-100% cross-slope). The design standards are different as per the type of terrain. For a given design speed, the cost of construction will vary as per the percent cross-slope of the terrain and steepness of the gradient.

• Vehicle characteristics: Vehicular characteristics can significantly affect geometric design of the road. Different vehicular dimensions, type and weight of axle, and vehicle dynamics (speed, acceleration, etc.) increase difficulty for choosing design vehicle which significantly influences design aspects of pavement, radius of the curve, clearance width, etc.

• Traffic flow: Traffic flow changes with time throughout the 24 hours. Traffic flow is low during early morning (6-9 a.m.), afternoon (12-6 p.m.), and late night (12-6 a.m.) hours whereas traffic flow is high during late morning (9 a.m. to 12 p.m.) and evening (6-9 p.m.) hours. Traffic flow is medium to high during late evening (9-12 p.m.) hours mainly due to operation of trucks. It becomes difficult to design a road facility considering the fluctuation in traffic flow. The roadway cannot be designed for a very low traffic flow or a high traffic flow (by considering peak hour traffic). A roadway designed for low traffic flow will face traffic jam for most of the time whereas a roadway designed for high traffic flow will be uneconomical to design.

Therefore, an optimal traffic flow value will be selected as a design hourly volume by considering an appropriate trade-off between facility and cost of construction.

5.3.1. Road alignment

Alignment refers to the position of the central line of the roadway. Horizontal alignment consists of straight and curved paths and vertical alignment consists of level and gradients. Finalizing an alignment is crucial since it is highly challenging to change it once the road has been built because doing so would increase the cost of construction and affect nearby road infrastructure.

Transportation engineers should pay attention while fixing an alignment for a roadway. The fixed alignment is easy to construct and maintain. The alignment should be short and straight as possible amid two terminal stations, however few deviations are considered based on practicality of the site. The construction cost, maintenance cost, and operating cost of an alignment should be economical. From a construction and operation standpoint, the alignment should be safe, especially near embankments, cuttings, and slopes.

5.3.2. Cross-sectional elements

It is important to study cross-sectional elements of the road and their importance before moving to geometric design. Pavement surface characteristics, camber, width of carriageway, and kerbs are a few major aspects to be considered in this regard. The following subsections focus on explaining these aspects.

5.3.2.1. Pavement surface characteristics

Pavement surface characteristics govern the life of a vehicle, comfort, and safety of the driver. Road friction, pavement unevenness, light reflection, and drainage are the four important aspects of pavement surface characteristics.

Road friction is a critical factor while designing horizontal curves. Friction between vehicle tyres and pavement surface is one of the factors governing operating speed and stopping distance. Longitudinal friction influences skidding or slipping of the vehicle and lateral friction influences countering centrifugal force (along with superelevation) while driving through a horizontal curve. The load acting multiplied by a factor known as the coefficient of friction, abbreviated as f, produces the frictional force that develops between the tyres and the pavement surface. IRC (Indian Roads Congress) suggests coefficient of longitudinal friction in between 0.35-0.4 and coefficient of lateral friction as 0.15.

Pavement unevenness is the undulations on a pavement surface. An even longitudinal profile is always desirable for designing high speed corridors, good driving comfort, less fuel consumption, safety, and for reducing vehicle operating costs. Bump Integrator is used to determine the undulations on a pavement surface in terms of unevenness index. A cumulative measure of vertical undulations of the pavement surface per unit length of the road is unevenness index (Mathew, 2019). The serviceability of the road as per unevenness index is shown in Table 1 below.

Table 1. Type of road as per the unevenness index

Type of road	Unevenness index (mm/km)
Good	< 1500
Satisfactory	1500-2500
Bad	2500-3500
Uncomfortable	> 3500

Color and light reflecting characteristics of the road surface impact visibility of the drivers. The road surface should be visible at night and should cause less glare issues during summer, rainy, and winter seasons. Further, impermeable pavement surface is required to avert water seepage into the pavement layers and should be provided with minimum cross-slope for draining rainwater from the surface.

5.3.2.2. Camber

Camber is a cross-slope provided to a road surface in the transverse direction to drain off the rainwater from the road surface. The camber is provided by raising the middle portion of the straight road surface to drain off rainwater in either direction. There exist different types of camber such as parabolic camber, straight line camber, and combination of parabolic and straight-line camber as shown in Figure 65. The rate of camber is represented by 1 in n. This ratio indicates one unit of vertical change per n units of horizontal distance. Camber is also expressed in percentage as x% (x in 100).

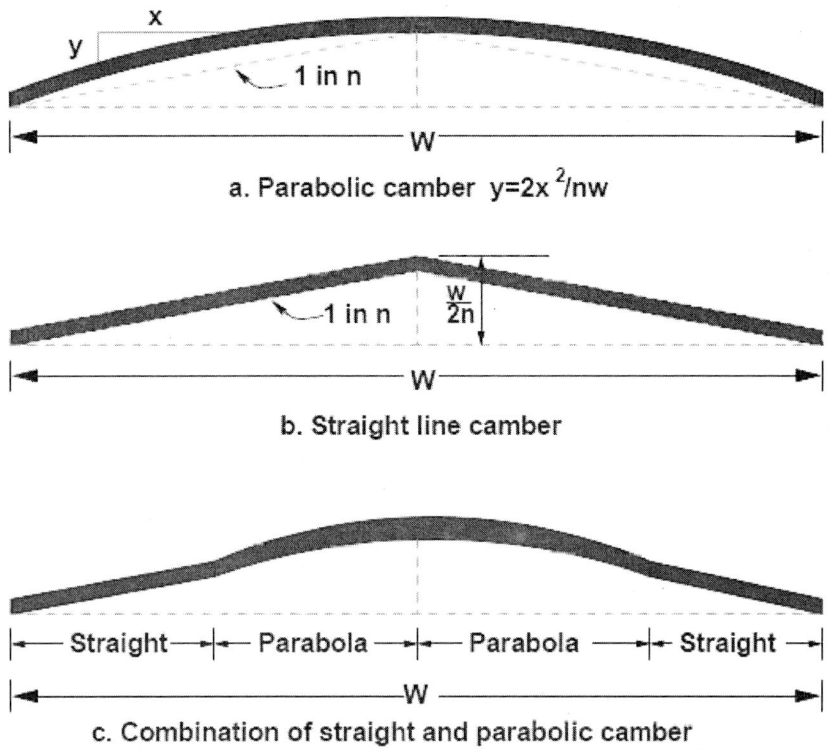

Figure 65: Different types of camber (Mathew, 2019)

5.3.2.3. Width of carriageway

Width of the carriageway varies according to the width of the traffic lane and total number of lanes in the carriageway. The width of the traffic lane is decided based on vehicle width and lateral clearance. The maximum allowable width of a vehicle is 2.44 m and desirable side clearance for single lane traffic is 0.68 m. IRC suggests a minimum lane width of 3.75 m for a single lane road. The minimum lateral clearance required is around 0.53 m on either side and 1.06 m in the center. Therefore, 3.5 m lane width is provided for a two-lane road (Mathew, 2019).

5.3.2.4. Kerbs

Kerbs specifies the margin between the road surface and the median or footpath or island or shoulder. Three different types of kerbs are low or mountable kerbs, semi-barrier type kerbs, and barrier type kerbs as shown in Figure 66. Low kerbs are provided to restrict traffic encroachment in other lanes by encouraging drivers to drive in the through lanes. The height of a kerb is about 10 cm above the pavement edge with a slope to enable vehicles to pass through the kerb and enter shoulder area at slow speed. Semi-barrier type kerbs and barrier type kerbs are provided when the pedestrian traffic is high. Vehicles can enter semi-barrier type kerbs with some difficulty whereas barrier type kerbs totally restrict the entry of vehicles.

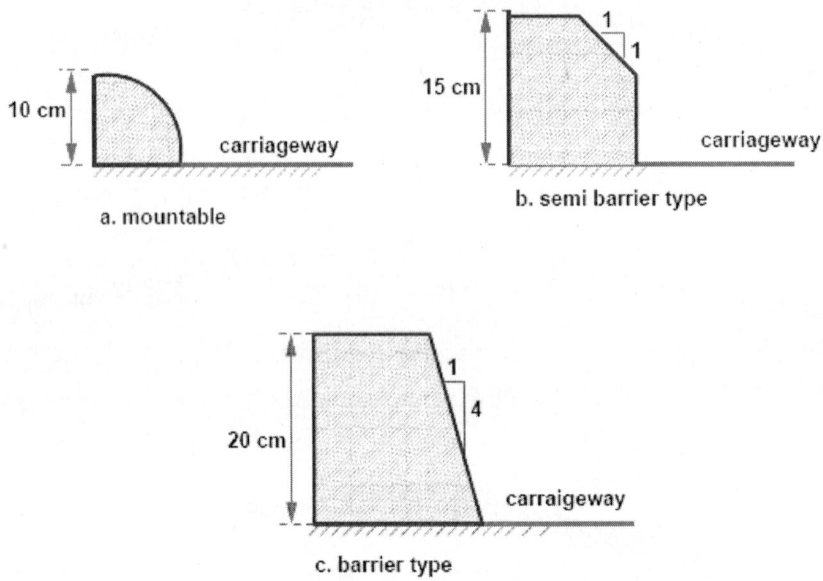

Figure 66: Different types of kerbs (Mathew, 2019)

5.3.3. Sight distance

The drivers require adequate sight distance to maneuver straight and curved roads as well as intersections. Sight distance mainly depends on reaction time, driving speed, brake efficiency, friction, and gradient of the road (Kadiyali, 2013). The reaction time is the time taken by the driver to detect the object to the instant when the brakes are applied (Pawar and Velaga, 2020). IRC recommends 2.5 seconds of reaction time for design purposes. Driving speed significantly affects sight distance. The drivers require more time to stop the vehicle while driving at high speed as compared to the drivers driving at low speed. Therefore, it is important to provide more sight distance at high-speed corridors. Brake efficiency of a vehicle is the braking force produced as a percentage of the total weight of the vehicle. Realistically, 100% brake efficiency cannot be achieved while braking. Thus, 50% brake efficiency is assumed for safe geometric design (Kadiyali, 2013). Road friction plays a prominent role in stopping the vehicle. High longitudinal friction will offer more resistance in between tyres and road surface leading to early stoppage of the vehicle. In this case, sight distance required to be provided will be less. Gradient is the rate of rise or fall along the length of the road. The drivers can easily stop their vehicle while driving up a gradient and thus the sight distance required is less. The drivers require more time to stop a vehicle while driving down a gradient. Because of this reason, the drivers need more sight distance to stop a vehicle.

The basic sight distances are:

- Stopping sight distance or the absolute minimum sight distance
- Overtaking sight distance
- Intermediate sight distance

5.3.3.1. Stopping Sight Distance

Stopping Sight Distance (SSD) is the minimum distance available to the driver on a roadway at any point of time to safely stop a vehicle. It is important to provide minimum sight distance equal to SSD to facilitate safe stoppage of a vehicle without any collision. The SSD is calculated by taking the sum of lag distance and braking distance. Lag distance is the length of road covered by a driver during reaction time t and braking distance is the length of road covered by a driver while braking to stop the vehicle.

$$\text{Lag distance} = v * t \tag{1}$$

$$\text{Braking distance} = \frac{v^2}{2gf} \tag{2}$$

$$\text{SSD} = \text{lag distance} + \text{braking distance} \tag{3}$$

Where, v is the velocity in m/s, t is reaction time in seconds, g is gravity (9.81 m/s2), and f is the coefficient of longitudinal friction (0.35-0.4).

5.3.3.2. Overtaking sight distance

Overtaking Sight Distance (OSD) is the minimum distance visible to the driver intending to overtake a slow-moving lead vehicle against the traffic in the opposite lane. OSD is measured along the center line of the road where the driver with his/her eye level of 1.2 m above the road surface is expected to detect an object which is 1.2 m above the road surface. OSD will be influenced by the speed of the overtaking vehicle, overtaken vehicle, and the vehicle traveling in the opposite direction. Further, spacing between the successive vehicles (vehicle attempting to overtake and the lead vehicle), reaction time, rate of acceleration of the driver intending to overtake, and gradient of the road will also influence OSD. The minimum OSD required on a two-lane two-way roadway can be estimated by using the equation below:

$$OSD = d_1 + d_2 + d_3 \tag{4}$$

$$d_1 = v_b * t \tag{5}$$

$$d_2 = 2s + v_b * T \tag{6}$$

$$T = \sqrt{\frac{4s}{a}}$$

$$s = 0.7v_b + 6 \tag{7}$$

$$d_3 = v * T \tag{8}$$

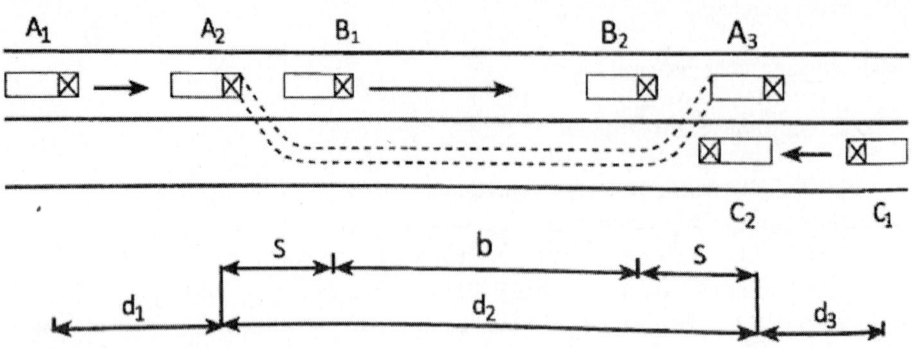

Figure 67: Overtaking sight distance (Chhabra, 2017)

Where, d_1 is the distance travelled by the overtaking vehicle A during reaction time t at speed v_b (m/s). d_2 is the distance travelled by the vehicle A while overtaking vehicle B in time T. d_3 is the distance travelled by the oncoming vehicle C at speed v (m/s) during time T as shown in Figure 67.

It can be noted that d_3 is only considered during undivided roads. d_3 will not be considered during divided roads.

Overtaking zones

Overtaking zones are provided when it is not possible to provide OSD for the entire roadway. Overtaking zones are the road sections dedicated for overtaking operations which will be indicated with road markings. Overtaking zones will be provided in the range of 3 times OSD (minimum requirement) to 5 times OSD (desirable length).

Intermediate Sight Distance

Intermediate Sight Distance (ISD) is provided when it is not possible to provide OSD. ISD is provided on horizontal curves or in restricted areas. ISD is twice the SSD.

5.3.4. Horizontal curve

A vehicle experiences outward force acting on it while negotiating a horizontal curve. This outward force is the centrifugal force which depends on speed of the vehicle and radius of the horizontal curve. The centrifugal force is countered up to some extent by the lateral friction between tyre and road surface. The centrifugal force tends to overturn the vehicle or slide outward from the center of the road curvature. The centrifugal force P (kg/m²) acting on the vehicle is represented as:

$$P = \frac{Wv^2}{gR} \tag{9}$$

Where, W is the total weight of the vehicle in kg, v is the vehicle speed in m/s, g is the acceleration due to gravity, and R is the radius of curve in m.

As mentioned above, centrifugal force may either overturn the vehicle about outer wheels or lead to transverse skidding.

At equilibrium, overturning is possible if,

$$\frac{b}{2h} = \frac{v^2}{gR} \tag{10}$$

The vehicle will be safe from overturning if,

$$\frac{b}{2h} > \frac{v^2}{gR} \tag{11}$$

Where b is the breadth of the vehicle and h is the height of center of gravity.

At equilibrium, skidding occurs when,

$$\frac{P}{W} = f = \frac{v^2}{gR} \tag{12}$$

The vehicle will be safe from skidding if,

$$f > \frac{v^2}{gR} \tag{13}$$

Where f is the lateral friction.

5.3.4.1. Superelevation

Superelevation is the transverse slope provided at the horizontal curve to counter the centrifugal force by raising the outer edge of the pavement with respect to the inner edge throughout the horizontal curve (Kadiyali, 2013). It is important to determine the required raise in the outer edge to have safe maneuver without toppling of the vehicle. The forces acting on the vehicle while maneuvering through the horizontal curve are shown in the Figure 68.

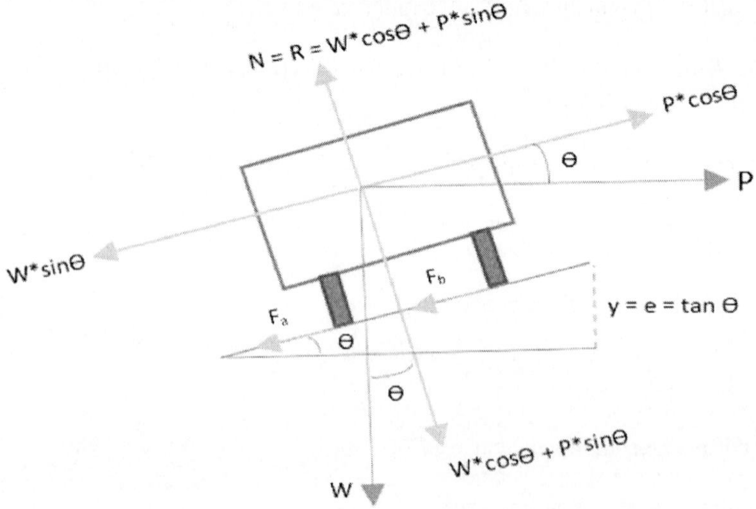

Figure 68: Forces acting on the vehicle while maneuvering through the horizontal curve (APSEd, 2022)

P is the centrifugal force, W is the weight of the vehicle, F is the lateral friction between tyre and road surface, and e is the rate of superelevation.

After equating all the forces, the expression for superelevation is as presented below:

$$e + f = \frac{v^2}{gR} \tag{14}$$

5.3.4.2. Design of other factors

Radius of horizontal curve and extra widening are two important components which are required to be designed other than superelevation for horizontal curves. The maximum comfortable speed depends on the radius of the curve. A ruling minimum radius R_{ruling} can be provided by assuming maximum rate of superelevation and coefficient of friction f as shown below:

$$R_{ruling} = \frac{v^2}{g(e+f)} \tag{15}$$

Preferably, the radius of the curve should be higher than R_{ruling}.

Extra widening is provided because a vehicle requires additional width while driving through a horizontal curve since rear wheels follow a path of shorter radius than front wheels as shown in Figure 69. Further, the drivers tend to drive away from the edge of the carriageway on a horizontal curve. Thus, extra widening required due to off-tracking (case 1) is called mechanical widening and extra widening required due to psychological constraints is called psychological widening.

$$\text{Extra widening} = \text{mechanical widening} + \text{psychological widening} \tag{16}$$

$$Mechanical\ widening\ W_m = \frac{nl^2}{2R} \tag{17}$$

$$Psychological\ widening\ W_p = \frac{v}{2.64 * \sqrt{R}} \tag{18}$$

Where, n is the total number of lanes, l is the length of the wheelbase, R is the radius of curve, V is the speed of the vehicle.

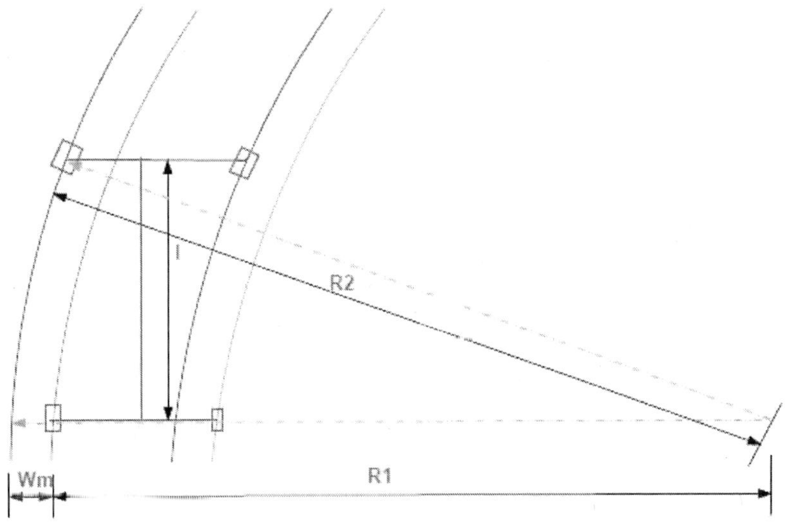

Figure 69: Extra widening (Mathew, 2006)

5.3.5. Transition curve

Transition curve is provided for gradually introducing vehicles to a circular curve directly from the straight road (Kadiyali, 2013). This is done to avoid sudden introduction of vehicles to centrifugal force which will cause shock and sway. A transition curve is provided before the circular curve having radius starting from infinity at the end of the straight road to gradually reducing to radius equal to circular curve. The objectives behind providing transition curve before circular curve are:

- gradual introduction of centrifugal force,
- gradual introduction of superelevation and extra widening,
- enable drivers to steer gradually through the curvature according to his/her own comfort and safety, and
- enhance the aesthetics of the road.

5.3.5.1. Types of transition curve

The different types of transition curve which can be provided are spiral (or clothoid), cubic parabola, and Lemniscate as shown in Figure 70. IRC recommends a spiral transition curve because the rate of change of centrifugal acceleration is consistent, and the radius of the transition curve is infinity at the end of the straight road and gradually changes to radius R at the start of the circular curve. Further, the calculation and field implementation of the transition curve is also very easy.

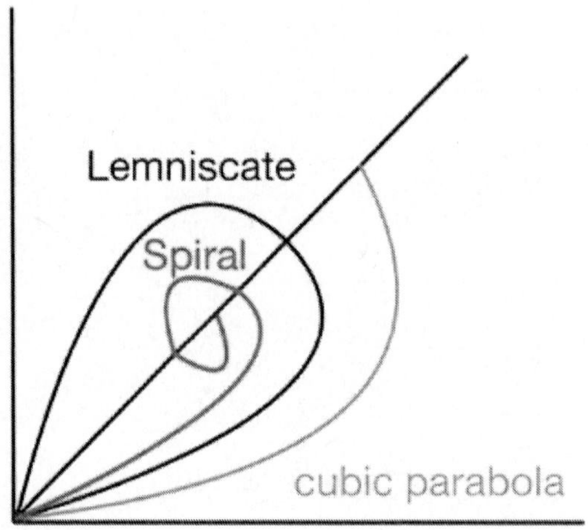

Figure 70: Types of transition curves (ESE notes, 2022)

5.3.6. Vertical curve

The vertical alignment of the road consists of gradients and vertical curves. Gradient is the rate of rise or fall along the road with respect to horizontal surface. The gradient is represented as 1 vertical to N horizontal. The gradient can be categorized as ruling gradient, limiting gradient, exceptional gradient, and minimum gradient. Ruling gradient (or design gradient) is the maximum gradient with which the vertical profile of the road is designed. India has different categories of roads because of which IRC has recommended certain values for ruling gradients according to types of terrain. Limiting gradient is considered when ruling gradient is not financially possible to construct. Exceptional gradient is a very steep gradient which is not usually considered except for some unavoidable situations. Exceptional gradient is limited for very short stretches. Minimum gradient is provided for drainage purposes. Camber is provided for lateral drainage and minimum gradient is provided for longitudinal drainage (Kadiyali, 2013; Mathew, 2006).

5.3.6.1. Grade compensation

An increased resistance to friction due to both gradient and curve will be experienced by the vehicle when a horizontal curve is in combination with a gradient. During such an instance, the gradient, if required, is reduced for the loss of tractive effort due to curve resistance. This reduction in gradient at the horizontal curve is called grade compensation.

IRC specifies criteria for the grade compensation.

I. Grade compensation is not necessary for the gradient within 4%.

II. Grade compensation is determined by estimating ((30+R)/R) %, where R is the radius of the horizontal curve in meters.

III. The maximum grade compensation is limited to (75/R) %

5.3.6.2. Summit curve

Summit curve is a vertical curve having convexity upwards. There is no requirement for a transition curve for introducing a summit curve because centrifugal acceleration is acting upward and not acting laterally outward or inward. A summit curve is formed when two gradients meet as shown in Figure 71.

- Positive gradient connects with another positive gradient (Figure 71 a)
- Positive gradient connects with a flat gradient (Figure 71b)
- Ascending gradient connects with a descending gradient (Figure 71c)
- Descending gradient connects with another descending gradient (Figure 71d)

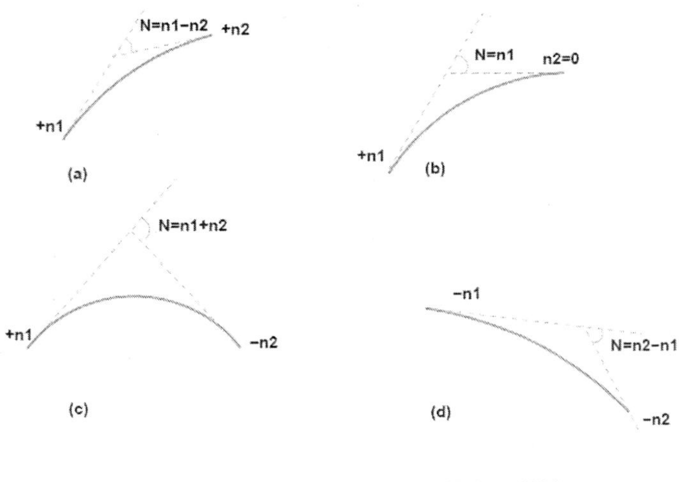

Figure 71: Types of Summit curves (Mathew, 2006)

5.3.6.3. Valley curve

Valley curve is a vertical curve having convexity downwards. They are formed when two gradients meet as shown in Figure 72.

- Descending gradient connects with another descending gradient (Figure 72a)
- Descending gradient connects with a flat gradient (Figure 72b)
- Ascending gradient connects with an ascending gradient (Figure 72c)
- Ascending gradient connects with another ascending gradient (Figure 72d)

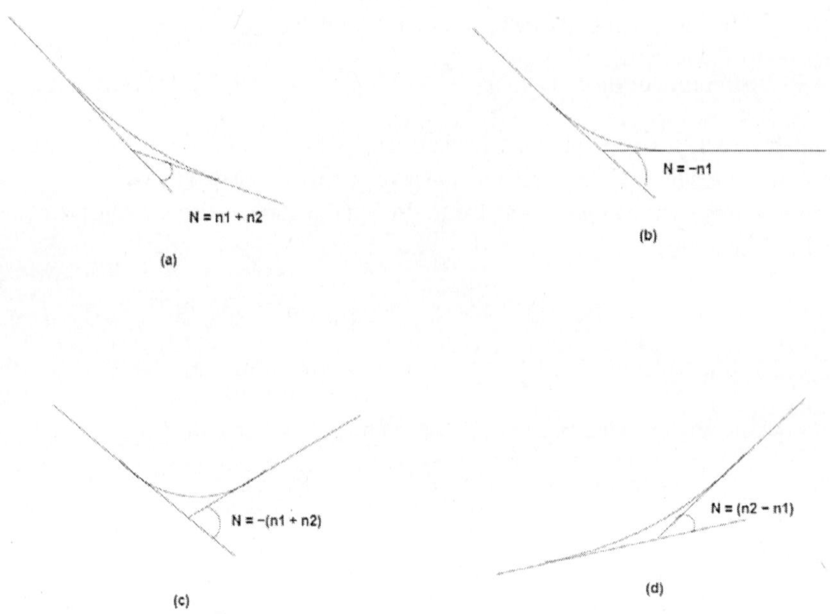

Figure 72: Type of valley curves (Mathew, 2006)

5.4. Pavement materials, design, construction, and management

5.4.1. Pavement materials

Pavement is a composite of different materials. Therefore, it is necessary to understand the behavior of the materials on an individual basis as well as in combination with other materials. Soil, aggregate, and bitumen are the three basic materials used to construct a pavement. A thorough understanding of these material properties is required since they affect stability, durability, and binding of the pavement structure.

5.4.1.1. Soil

Soil is the main material for the construction of embankments and subgrade of highways. Soil is also used in other pavement layers such as base and sub-base courses (Mathew, 2006). The pavement is constructed over a subgrade which supports the pavement layers. The performance of the pavement, especially flexible pavement majorly depends on the subgrade soil and its properties. Various laboratory tests are developed to explore physical properties and soil strength. A wide diversity of soil can be observed throughout and thus, it is desirable to classify the subgrade soil according to the similar physical properties. Grain size analysis and consistency tests are performed to classify subgrade soil according to similar physical properties (Khanna et al., 2013).

Soil compaction is an important phenomenon where the soil particles are constrained to be packed together by decreasing air voids which reduces chances of future settlement of earth embankment. Soil compaction is conducted to reduce the air voids and increase dry density to increase shearing strength. A compacted soil fill and subgrade improve load bearing ability of the pavement which results in reduction of pavement thickness. (Khanna et al., 2013).

Different tests are developed for determining soil strength. Tests conducted on soil to evaluate strength properties can be divided into three groups: (a) shear tests, (b) bearing tests, and (c) penetration tests. California Bearing Ratio (CBR) test is a well-known test conducted for evaluating the strength of subgrade soil for the design of flexible pavements. CBR test denotes the resistance offered by the soil to penetration using a standard plunger under controlled conditions (Khanna et al., 2013).

Soil sometimes is not directly suitable for use in subgrade as a pavement layer. However, it is possible to improve the strength and durability of the soil by adopting appropriate soil stabilization techniques. Soil is mostly stabilized using mixed Portland cement in appropriate proportion since it is fairly suitable for various soil types. Cement stabilized soils are commonly used in subgrade, subbase, and in base course layers of some pavements (Khanna et al., 2013).

5.4.1.2. Aggregate

Stone aggregates are widely used in the construction of different pavement layers forming one of the prime materials of the pavement structure. Aggregate accounts for 92% – 96% of bituminous concrete and around 70% – 80% of Portland cement concrete (Mathew, 2006). Aggregates used in the surface course have to mainly bear load stresses occurring on the road, withstand the impact due to heavy wheel loads, and have to resist abrasive action of traffic movement under all weather conditions. Thus, the aggregates should be hard, possess resistance to crushing, and need to have adequate resistance to abrasion offered by the traffic movements. The aggregates should offer resistance to getting smooth or from getting polished due to fast and rapid movement of vehicles to prevent the pavement surface becoming slippery which can result in skidding related accidents, particularly during monsoon season. The aggregates should not easily disintegrate under adverse weather conditions and offer resistance to weathering action. Determination of specific gravity of the aggregates is important to assess strength and suitability of the aggregates. The aggregates should not have high air voids or pores which will result in low specific gravity indicating low strength and durability of the aggregates. The aggregates which will be used for construction should not be too flaky or elongated because they get easily crushed during compaction and get break down under heavy traffic loads. Angular aggregates provide better stability due to interlocking which is desirable for flexible pavements. Finally, affinity of the aggregates to bituminous binders is an important property signifying good adhesion to bituminous materials in water. The pavement layer will be in contact with water for prolonged periods during monsoon season. The bituminous binder must not get stripped from the coated aggregates if they have affinity to the bituminous binder (Khanna et al., 2013; Mathew, 2006). The desirable properties of the aggregates are:

- Resistance to crushing (aggregate crushing test)
- Resistance to impact or toughness (aggregate impact test)
- Resistance to abrasion or hardness (Los Angeles abrasion test)
- Resistance from getting smooth or polished (polished stone value test)
- Resistance to soundness or weathering (soundness test)
- Good adhesion with bitumen materials in presence of water (stripping value test)
- Good shape factor to avoid too flaky and elongated particles of coarse aggregates (shape test)

5.4.1.3. Bitumen

Bitumen is a petroleum product obtained by the distillation of petroleum crude used in pavement construction works (Khanna et al., 2014, 2013; Mathew, 2006). The bitumen is brought to an adequate fluidity or viscosity before use in pavement construction by either heating in the form of hot bitumen binder or by dissolving in light oils in the form of cutback bitumen or by dispersing bitumen in water in the form of bitumen emulsion. Bitumen binders are commonly used in surface courses of the pavement. They are used in a base course of flexible pavement to withstand adverse traffic and climate conditions. Bitumen is also used to stabilize the soil and to prepare sealer materials for filling joints in cement concrete pavements. Viscosity test, penetration test, ductility test, and softening test are the common tests used to evaluate properties of the bitumen (Khanna et al., 2014, 2013; Mathew, 2006).

5.4.1.4. Bituminous mixture

Surface course of the flexible pavement needs to withstand high stress conditions, wear, and tear due to traffic loads, and adverse climatic conditions. Therefore, a high quality hot bituminous mix is required to be designed and laid on the surface course of the flexible pavement to sustain heavy traffic loads, wear, and tear due to high-speed traffic movement and provide adequate skid resistance in wet conditions. Bituminous mix consists of well graded hard aggregates and suitable bituminous binder at correct proportion. High-quality bituminous mixtures must be able to withstand repeated stresses and deformation, and resultant fatigue. Additionally, it must be sufficiently resistant to moisture-induced damage, low-temperature cracking, and resistance to permanent deformation during hot weather conditions. The workability of the mix should be adequate at the mixing, laying, and compacting temperatures. A variety of laboratory experiments are developed for examining bituminous mixtures. Marshall Stability test is most commonly used for evaluating the adequacy of a bituminous mix (Khanna et al., 2014, 2013; Mathew, 2006).

5.4.2. Pavement design and construction

Pavement consists of overlaid layers above the subgrade soil designed to distribute the traffic load to the subgrade. The pavement surface must offer appropriate driving quality, adequate skid resistance, favorable light reflecting characteristics, and low noise pollution (Mathew, 2006). The main purpose of the pavement design is to ensure that the transferred stress due to the traffic load should be within the bearing capacity of the subgrade. Flexible and rigid pavements are the two different pavement structures mainly designed according to the requirement. The following subsections provide an overview of the same.

5.4.2.1. Flexible pavement

Flexible pavement is a road structure consisting of different layers designed to transmit load to the subgrade soil. Wheel load stresses in flexible pavement are transferred by grain-to-grain contact of the aggregate through the granular structure as shown in Figure 73. The wheel load acting on the pavement gets distributed over a wider area because of which stress reduces with the depth. For this reason, flexible pavements consist of various layers for transmitting wheel load stresses. Flexible pavements have low flexural strength and are rather flexible in their structural action under the loads (Khanna et al., 2014). The deformation in the layers of flexible pavement may be recoverable or non-recoverable (Mathew, 2006). Thus, if a lower layer or subgrade gets deformed due to permanent deformation then the upper layers will also reflect similar deformation.

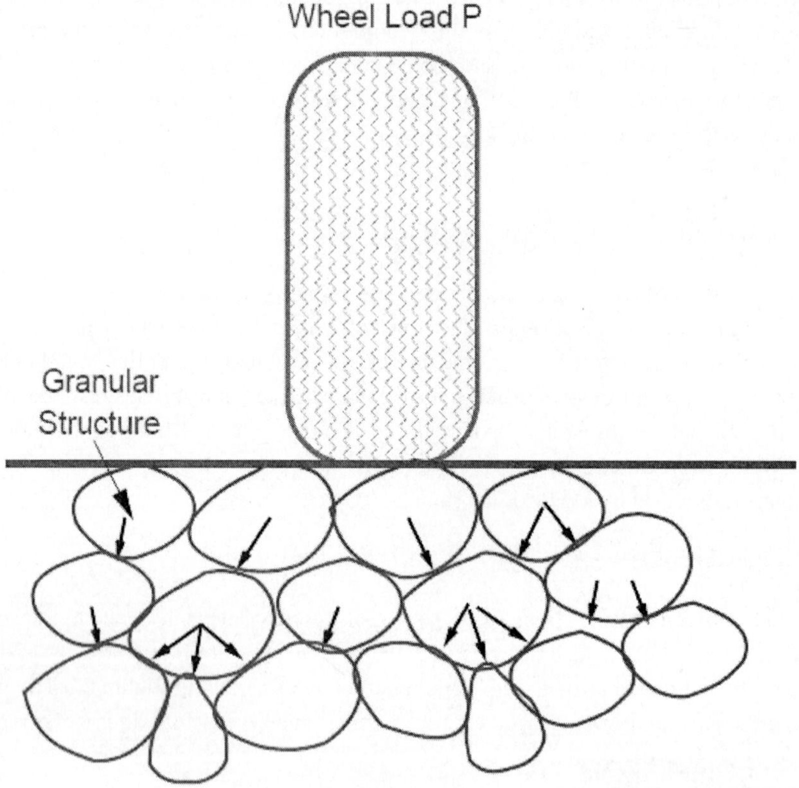

Figure 73: Load transfer in granular structure (Mathew, 2006)

A typical flexible pavement consists of a subgrade, sub-base, base, and surface course as shown in Figure 74. The top 500 mm of embankment consists of subgrade, which is prepared to bear the stresses transmitted from the above layers. The subgrade soil should be adequately compacted to the optimum moisture content for achieving the desired dry density for limiting the scope of rutting during the service life of the pavement. The sub-base course is the layer above the subgrade whose primary function is to provide structural support, improve drainage conditions, and decrease incursion of fine particles from the subgrade in the pavement structure. The base course is the layer above the sub-base course provided for sustaining wheel load stresses and for sub-surface drainage. The pavement surface has to bear maximum vertical compressive stress equal to contact pressure under the wheel load. Therefore, the top layer (i.e., surface course) has to be of the best quality which can sustain the vertical compressive stresses and wear and tear due to continuous movement of traffic and weather conditions (Khanna et al., 2014, 2013; Mathew, 2006).

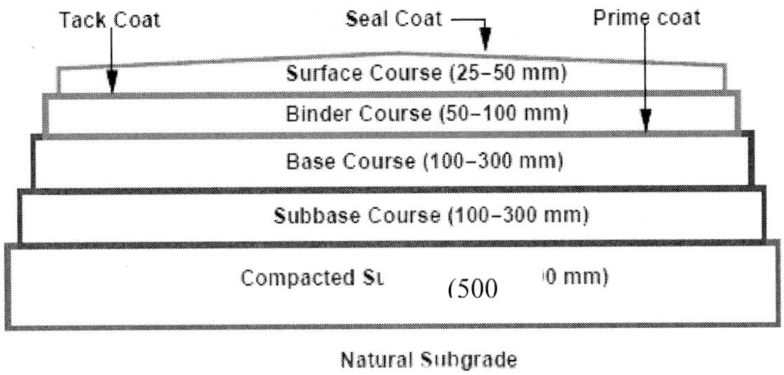

Figure 74: Typical cross-section of flexible pavement (Mathew, 2006)

5.4.2.2. Rigid pavement

Rigid pavements are made of Portland Cement Concrete (CC). Figure 75 shows the cross-section of rigid pavement. Plain cement concrete slabs designed for particular strength characteristics are laid on the subgrade with or without steel reinforcements at the joints. The rigid pavement slabs constructed using high quality plain cement concrete are expected to sustain 45 kg/cm^2 of flexural stresses. A subgrade or a single layer of stabilized or granular material is generally prepared on which pavement slabs are placed for constructing a rigid pavement structure. The rigid pavement transfers the wheel load stresses through 'slab action' to the much wider area beneath the pavement slab. The pavement behaves like an elastic plate over a viscous foundation as shown in Figure 76. The CC slab is subjected to tensile stresses based on the bending location of the slab under wheel load and temperature differences.

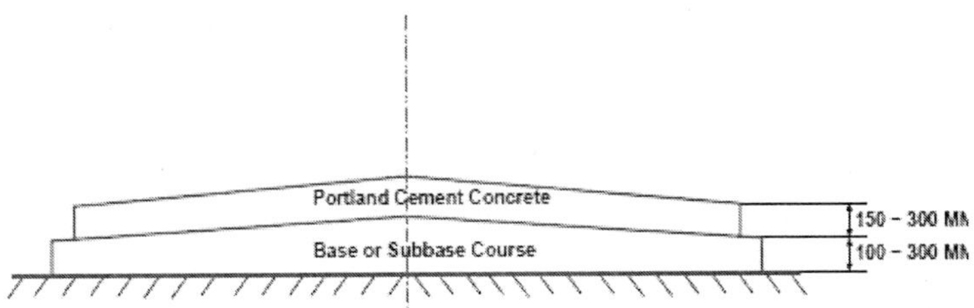

Figure 75: Typical cross-section of rigid pavement (Mathew, 2006)

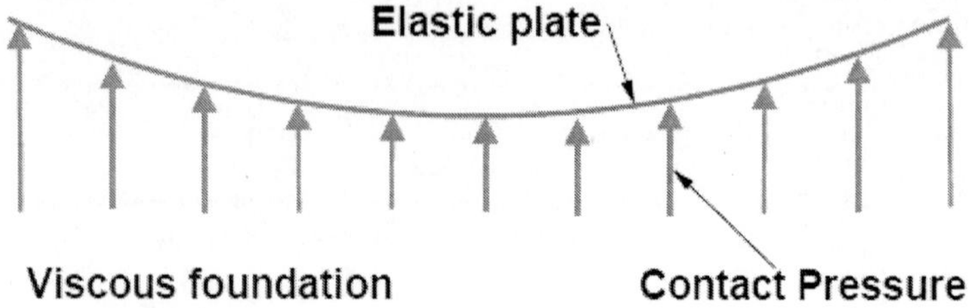

Figure 76: Concept of elastic plate on viscous foundation (Mathew, 2006)

The pavement slab made of high-quality cement concrete serves as a wearing surface as well as base course. The base or sub-base course is laid under the CC pavement slab by providing a good drainage facility (as a layer) which will increase the life of the pavement (Khanna et al., 2014). Therefore, a typical rigid pavement consists of (bottom layer to top layer) subgrade soil, drainage layer, lean cement concrete or dry lean concrete as a base course or a good granular sub-base course, and CC pavement slab (Khanna et al., 2014).

5.4.3. Pavement management

Road construction is an important component in infrastructure development and road, or pavement maintenance is important for development of the country without any hindrance. Pavement maintenance is a vital element of the pavement system. It is important to assess the functional and structural condition of the pavement for diagnosing the problem in order to either preserve or replace the existing surface to maintain the strength of the pavement. Pavement starts deteriorating from the day they are opened for traffic due to the collective effect of environment and traffic conditions. It is very important that each layer of the pavement has sufficient stability for withstanding the traffic under varying environmental conditions. The pavement is constructed for particular design traffic. However, it is likely that the actual traffic volume increases beyond the original design capacity which leads to pavement deterioration and distress and decreasing the design life of the pavement. Therefore, it is very important to conduct regular pavement evaluation at certain time intervals to determine the extent of pavement degradation so that appropriate maintenance measures can be adopted to improve strength and stability of the pavement with suitable overlays at the right time. Implementing pavement maintenance and adopting appropriate pavement treatment at the right time is the important component of the pavement management system which results in overall savings by minimizing life-cycle cost of the pavement.

Functional assessment of the pavement surface is important to decide the appropriate maintenance measure including the type of resurfacing required on the identified sub-stretches. Structural evaluation helps in assessment of structural adequacy of the pavement in sustaining prevailing traffic, weather, and environmental conditions. Based on functional and structural evaluation, overlay thickness is designed to strengthen the pavement structure and prolong the life up to the desired period. Delay or failure in carrying out the pavement maintenance to strengthen the pavement structure will result in loss in structural capacity, rapid rate of deterioration and failure of the pavement structure during the prevailing period. Further, delay in pavement maintenance results in increased maintenance cost and leads to expensive rehabilitation treatment of more than one pavement layer for restoring the strength of the pavement structure or may require re-construction of the identified road stretches.

A wide range of methods and techniques are available for structural and functional evaluation of flexible pavements. Benkelman beam rebound deflection and falling weight deflectometer are most commonly used for structural evaluation of the flexible pavement. IRC has provided guidelines for conducting Benkelman beam rebound deflection study to design the overlay thickness required to strengthen flexible pavement. Benkelman beam rebound deflection is easy to conduct and is therefore generally conducted for structural evaluation of the flexible pavement.

5.5. Public transport systems

In general, public transport is commonly referred to as land-based passenger transport via bus or train. In the current scenario, most of the population adopts a private mode of transport (personal two-wheeler or car) instead of public transport. Private transport is mainly chosen due to comfort, convenience, quick travel, no constraint of bus or train schedule, and privacy. In today's world owning a two-wheeler or a car is comparatively easier than a few years ago. Because of this reason, vehicle ownership is increasing day by day and the roads are getting crowded. It is difficult to widen a particular road because of many constraints. Therefore, it is necessary to develop and promote the use of public transport among the daily commuters for reducing traffic congestion and air as well as noise pollution.

5.5.1. Development of public transport system

An efficient public transport system can be developed only by means of careful planning. Public transport within urban areas is dominated by bus service. It is very important to conduct travel surveys for selecting bus routes which will cover major locations along the route in one trip. Passengers will not be willing to frequently transfer from the start of their travel to reach their desired destination. Therefore, the route and the bus-stops must be properly planned so that the walking distance should not be too high which will not involve more than 10 minutes of walking (Kadiyali, 2013). Further, the bus service frequency should be managed in accordance with the passenger demand. This means that high frequency of buses can be provided during peak working hours whereas less buses can be operated during non-peak hours. The comfort and safety of the passengers (seating as well as standing) in a bus must be given special attention. It is very important to plan roads in urban areas for bus traffic. A provision of bus-bay is very important for separating bus stoppage from the carriageway. A bus stopping in a carriageway will create a temporary bottleneck which will significantly reduce the capacity of the road thereby deteriorating the overall traffic operation. Moreover, a minimum of 75 m of distance is required to be maintained between a bus stop and intersection for the safety of the passengers (CRRI, 2017). Exclusive bus lanes can also be dedicated for operating buses to reduce journey time by exempting them to travel from the regular traffic. Finally, parking spaces can also be provided at major bus stations so that passengers can travel to the bus station with their private vehicle, can park their vehicle, and use bus service for reaching his/her desired destination.

5.5.2. Promotion of public transport

The government agencies should promote public transport so that people start prioritizing the public transportation system over private vehicles. This can be done by integrating different modes of public transportation, usage of modern technology, smart fare management, and improving user experience. Intermodal public transportation can revolutionize the public transport system by providing easy, hassle-free, and convenient modes of transport as per the requirement and type of journey. Different modes of transport can be appropriately combined which can offer door-to-door connectivity options. For example, metro as well as bus service can directly provide access to the railway station or an airport where the passengers will not have to struggle to reach the final destination using a private vehicle or through intermediate public transport (like auto rickshaws). Public transport systems can adopt modern technologies like electronic-ticketing, fleet management, real-time monitoring and tracking of buses, dynamic passenger information system, etc. for improving passenger safety and fleet efficiency which will enhance the overall user experience.

5.5.3. Indian public transportation

India has observed rapid growth in the number of motor vehicles in the last two-to-three decades. This rapid growth of private vehicles has escalated the congestion problems leading to higher travel times in most parts of the cities. Therefore, there is an urgent requirement to plan the transportation systems and facilitate dedicated space for public modes for promoting public transportation. This can be achieved by developing a mobility plan for each city by giving importance to various public transportation services like BRTS (Bus Rapid Transit Service), metro, monorail, etc. Various cities in India are productively operating rapid public transportation services. Cities like Mumbai and Delhi have multiple public transport services like rapid bus transit and rail transport services for the commuters. Ahmedabad BRTS service as shown in Figure 77 is considered as one of the best bus public transport services of India (interested readers are directed to Smart Cities Council (2016) for detailed information). Mysore city has significantly improved the efficiency of bus services by incorporating ITS for operating and managing the fleet of about 400 buses from 3 depots within the city (interested readers are directed to The World Bank (2011) for detailed information).Figure 78 shows the online bus tracking platform available on the Mysore ITS.

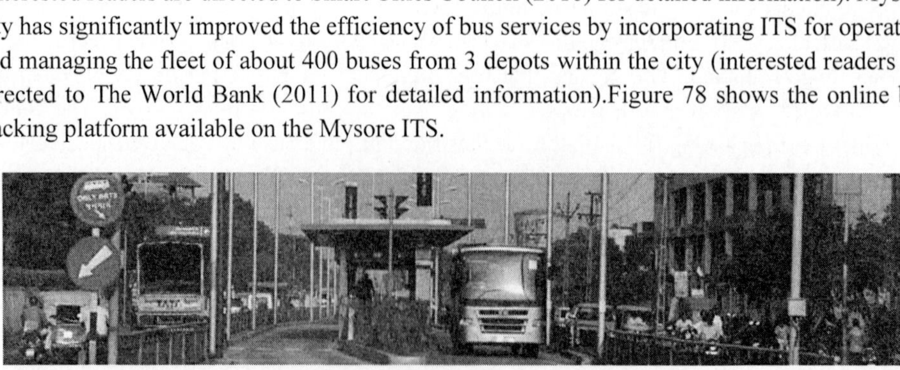

Figure 77: Ahmedabad BRTS

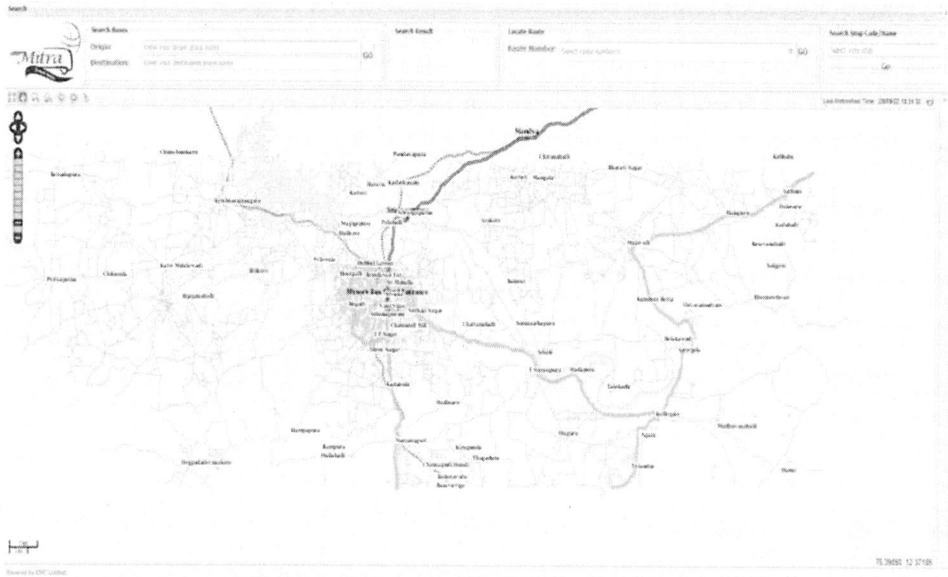

Figure 78: Mysore Intelligent Transport System (Karnataka State Road Transport Corporation, 2016)

5.6. Transportation logistics

5.6.1. What is logistics?

Planning and carrying out the efficient movement and storage of commodities from their point of origin to their site of consumption is the process of logistics (Essex, 2022). Logistics involves management of resources from acquisition, storage, to transportation. Logistics was initially used to explain the procurement and movement of military equipment and supplies. Currently, logistics is closely associated with the global supply chain. The manufacturers or producers rely on logistics for maintaining efficiency at reduced costs to ensure to fulfil consumers' needs. Logistics is one of the fastest growing industries and is expected to witness a boom considering the increasing prominence of e-commerce in daily life. The main activities included in logistics are:

(i) Processing of the order

(ii) Inventory management

(iii) Warehouse management for commodities storage and packing

(iv) Efficient, cost-effective, and safe transportation

(v) Timely delivery of the commodities

5.6.2. Effect of transportation on logistics

Transportation is a key element which plays a connective role in management of logistics from production of a product to delivering it to the customer. Different modes of transportation are effectively used by properly planning the transfer of commodities from one place to another. Containerization is a major component in logistics and supply chain management where standardized containers are used for shipping the commodities as a single unit. Standardization of containers has facilitated intermodal transportation from the production unit to warehouse to customer delivery. The logistics service provider decides the exact combination of transportation modes based on the product to be shipped considering a cost-effective approach. Further, the use of information technology and ITS has changed the entire dynamics of the logistics industry. Up-to-date information to the customers starting from packing, shipping, to the delivery of the product has increased the reliability of the customers in the logistics industry. Further, ITS plays a prominent role in managing different modes of transportation, handling bulk cargo, and the continuous tracking of shipments using GPS or any other tracking device which helps in mitigating the problems associated with delivery services like dispatching of the product to the accurate destination.

5.7. Road safety

Road traffic safety is a growing concern playing a prominent role in a country's development by impacting the economy and public wellbeing. Road traffic accidents are the eighth most common cause of fatality globally (World Health Organization, 2018). More than 1.35 million fatalities are caused by road traffic accidents each year (World Health Organization, 2018). According to the report published by World Road Statistics (2018), India ranks first in road traffic fatalities across 199 countries followed by China. The road accident statistics of India showed that around 51 accidents and 17 fatalities occur every hour (Ministry of Road Transport & Highways, 2020). The vehicular population in India is increasing at a 10.11% compound annual growth rate (Ministry of Road Transport & Highways, 2020). The current road infrastructure is not able to adequately withstand the swift vehicular growth resulting in high vehicular interactions thereby elevating traffic safety concerns.

Road traffic accidents are complex and multifaceted events resulting due to interaction between numerous factors. These factors can be broadly classified as (a) human factors, (b) vehicle factors, and (c) road and environment factors. Human factors are responsible for more than 90% of road traffic accidents, with the remaining 10% being caused due to vehicles, road, and environment related factors (Soehodho, 2009). Human factors in transport safety can be attributed to all those factors that impact driver behavior during safety-critical situations. The road traffic accidents associated with vehicle characteristics can be accredited to the vehicle factors. The road traffic accidents caused due to the influence of geographical area, improper design of road infrastructure, and bad weather conditions can be related to road and environmental factors.

Road traffic safety is mainly evaluated using a reactive approach and a proactive approach (Pawar et al., 2022a). Road traffic safety is evaluated using a reactive approach by developing statistical models by obtaining historical accident data. Proactive approach analyses road traffic safety using different surrogate safety measures by evaluating near-accident events as alternatives to real accidents. Road traffic safety is also evaluated by conducting a road safety audit of a particular road section before, during, and after the road construction. The following subsections provide a brief overview of different aspects of road traffic safety.

5.7.1. Causes of road traffic accidents

5.7.1.1. Human factors

Driver, vehicle, road, and environment are the four main and basic factors governing road accidents. The majority of the road accidents are caused by variability in driving behavior and human errors. Human performance is very difficult to predict because it varies from individual to individual. The variability in performance and decision making of humans make the system very complex to deal with. The humans' ability to perceive, react, vision, hearing, and other physical, psychological, and physiological characteristics vary based on age, gender, driving experience, driving profession, marital status, influence of external and internal factors. Further, road accidents because of human errors are the product of traffic rules violations, driving without a valid driving license, and driving without safety precautions.

In recent years, traffic safety researchers observed that driving performance of the drivers was significantly influenced by numerous external factors like sleep deprivation, mobile phone distraction, time pressure, alcohol impaired driving, etc. (Choudhary and Velaga, 2018; Mahajan and Velaga, 2021; Pawar and Velaga, 2021; Yadav and Velaga, 2021). Mobile phone distraction resulted in drivers reacting 40% to 204% late as compared to normal driving (Choudhary and Velaga, 2017). Alcohol consumption while driving negatively affected brake transition time where drivers were observed to take more time to achieve maximum deceleration by 16% to 53% than normal driving (Yadav and Velaga, 2021). Speed variability of sleep deprived drivers was observed to increase by 1.28 to 1.34 times as compared to normal driving (Mahajan and Velaga, 2021). The drivers under time pressure were observed to drive quickly, accept short gaps at unsignalized intersections, and were more likely to cross a signalized intersection when the signal turned from green to yellow than normal driving (Pawar et al., 2022b; Pawar and Velaga, 2022). Through all these results, it is clear that driving performance of the drivers significantly deteriorated which apparently increased the likelihood of accidents.

Traffic rules violation is a serious road safety issue in India. Over-speeding is the most common problem followed by lane indiscipline leading to nearly 72% and 6% of the total road accidents, respectively caused due to traffic rules violations (Ministry of Road Transport & Highways, 2020). Drunken driving, use of mobile phones, and red-light jumping is also one of the major traffic rules violations observed in India leading to 2.7%, 2.3%, and 1% of the total road accidents caused due to traffic rules violations, respectively (Ministry of Road Transport & Highways, 2020).

Drivers without a valid driving license are a serious threat to the overall road users because they are not trained or skilled to drive through traffic and different hazardous events. Drivers without a valid driving license have low hazard perception skills as well as they are not trained to control the vehicle during difficult traffic events. The inability of the drivers without a valid driving license to control the vehicle jeopardizes safety of the other road users. Further, the drivers using two-wheelers and four-wheelers in daily life do not usually take safety precautions like using helmets and seatbelts. Usage of helmets and seatbelts does not prevent the occurrence of an accident but will play a crucial role in preventing fatal or grievous injury during road accidents. Therefore, drivers must practice using helmets and seatbelts while driving two-wheelers and four-wheelers for their own safety. These issues are very common in India which are required to be addressed quickly with strict enforcement from the local traffic authorities.

5.7.1.2. Vehicle factors

Age and overloading of vehicles are the two major vehicle factors associated with road accidents. MORTH report on road accidents in India also advocates the above mentioned two reasons as the major issues associated with road accidents. According to the MORTH report on road accidents, vehicles of age between 10-15 years, more than 15 years, and age not known accounted for 12.5%, 11%, and 14.6% of the total road accidents (Ministry of Road Transport & Highways, 2020). Road accidents increased for categories: vehicles with age above 15 years from 9.6% to 11% and vehicles age not known from 13.9% to 14.6% in 2019 over 2018 (Ministry of Road Transport & Highways, 2020).

Truck drivers in India generally overload containers for delivering more goods in order to earn more incentives. However, overloaded trucks are more prone to accidents because (i) of the wearing down of braking system, (ii) road collapses, (iii) bursting of tyres, (iv) roll-over caused by shift in the center of gravity, and (v) excess weight degrading working of trucks during uphill and downhill (Ministry of Road Transport & Highways, 2020). In 2019, India recorded 7.9% of the total accidents due to overloading of trucks from which 9.5% of accidents resulted in fatality and 8.2% of the drivers were injured (Ministry of Road Transport & Highways, 2020). Considering this, MORTH is taking strict actions against overloading of trucks which resulted in a decline in road accidents in 2019 as compared to 2018.

5.7.1.3. Road and environmental factors

Road characteristics and weather conditions have a significant influence on road accidents. Road alignment, sight distance, superelevation, roadway width, shoulders, road signs and markings, intersection design, road surface characteristics, and median width are the major road factors influencing safety of the drivers. Stopping sight distance is required for the safety of the drivers to detect the vehicles along horizontal and vertical curves. Superelevation is required to be provided for safely negotiating the horizontal curves. Deficiency of superelevation may compromise the safety of the drivers. A minimum carriageway of 7 m must be provided for a two-lane undivided road with bidirectional traffic and 5.5 m width of intermediate lane where two-lane road cannot be provided, and traffic cannot be negotiated through a single lane. Further, overtaking of vehicles should be strictly restricted on two-lane and intermediate lane roads.

Paved shoulders can be provided instead of earthen shoulders wherever possible with adequate widths because Indian drivers often use shoulders as an extra lane during traffic congestion. Road signs and markings play an important advisory role to the drivers at every step of driving. It is very important to provide clear information without creating any confusion through road signs and markings for safe maneuver of the drivers. Road or pavement surfaces constructed must be anti-skid even when the road surface is wet. This will facilitate emergency application of brakes during hazardous traffic events. Median width is of great importance since width of the median will determine the detrimental effects of glare on the opposite stream drivers by compromising their visibility.

Weather has a significant effect on road surface condition and visibility which substantially governs the driving behavior of the drivers. Road surface becomes wet during the rainy season, leading to slippery driving conditions which degrades skid resistance of the roads. During the winter season, snow or ice also makes the road surface slippery leading to less skid resistance from the roads. During winter as well as during early morning hours, fog can reduce visibility of the drivers. Thus, bad weather conditions significantly reduce visibility and skid resistance of the roads during which drivers need to slowly and carefully negotiate the road.

5.8. Advanced technologies and ITS/IT/IoT

5.8.1. What is ITS and why is it necessary?

ITS stands for Intelligent Transportation Systems which uses communication technology by deploying and integrating the information networks (road sensors) to improve traffic mobility, efficiency, and safety. The basic idea behind the development of ITS is to use modern technology by developing the digital road infrastructure along with physical road infrastructure to resolve a variety of road traffic problems like road congestion, re-routing of vehicles, toll collection, management of public transport systems, road pricing, etc. ITS links drivers, vehicles, and road infrastructure with each other for improving safety and capacity of road systems and reducing air and noise pollution.

5.8.2. Components of ITS

Location, mapping, and communication are the three most important components required for establishing ITS as shown in Figure 79. Global Positioning System (GPS) is a satellite-based navigation system which is the most commonly used technology for identifying locations. Mapping is a technology used to plot the location information on a digital map. Communication is a vital component of ITS which can be established via a variety of broadcasting and telecommunication technologies such as Dedicated Short-Range Communication (DSRC), mobile phones, etc. Further, an onboard unit is also required for establishing a common platform that is compatible with the developed digital infrastructure for utilizing ITS.

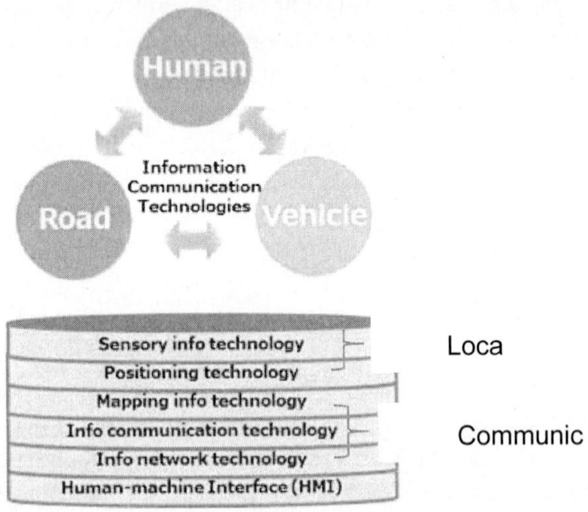

Figure 79: Components of ITS

5.8.2.1. ITS configuration

ITS can be configured as a (i) two-way communication system, (ii) one-way communication system, and (c) telematics system, among other things (Maniko et al., 2016). A two-way communication system is also known as an active system that can be readily deployed for a variety of applications through interactive communication. The one-way communication system is a passive system which requires an onboard unit for establishing communication with the digital road infrastructure. Telematics is a wireless network system used for gathering and exchanging data such as vehicle location, driving behavior, vehicle characteristics, etc. with the digital road infrastructure which is continuously evaluated along with the changing traffic conditions based on which decisions are taken for the effective use of the road network considering the safety of the transport systems.

Integration of the current road infrastructure with ITS is a challenging work which highly depends on the financial condition of the road users. Incorporation of ITS in the road system must be planned properly by focusing on the current financial condition and by taking into account the future scalability. A high technologically advanced system will not be beneficial if the road users are not willing to adopt the system whereas a low ITS system will get outdated in a short span which might put additional financial burden on the country in order to cope up with the ever-growing technology.

5.8.2.2. Communication

Development of ITS majorly depends on setting up different communication devices along the roadway. Different communication devices such as DSRC, GSM/GPRS, etc. are placed along the roadway which connect with the vehicles through an onboard unit. An onboard unit is a communication device mounted on a vehicle which acts like a bridge to establish communication with digital road infrastructure, transfer vehicle information, and perform various tasks such as payment of tolls, monitoring freight transport, providing information regarding efficient driving routes, etc. (Maniko et al., 2016). The onboard unit acquires information from multiple sensors and manages large data at high computation speed because of which it has become one of the important components for in-vehicle communication. The onboard unit is connected with GPS which monitors the movement of the vehicle. The onboard unit also gets connected with multiple road sensors like DSRC and GSM (Global System for Mobile)/GPRS (General Packet Radio Service) where the data is exchanged between the two systems (Maniko et al., 2016; Vaculík et al., 2008). The GSM/GPRS collects driving data through an onboard unit which is connected with the monitoring center through the internet. The data is analyzed and the necessary information is conveyed to the driver through the onboard unit. Figure 80 illustrates a typical example of communication between an onboard unit and the roadside sensors. The researchers working on the development of ITS are making efforts to integrate onboard units with V2V (Vehicle-to-Vehicle), V2P (Vehicle-to-Pedestrian), and V2I (Vehicle-to-Infrastructure) for enhancing safety of the overall road users.

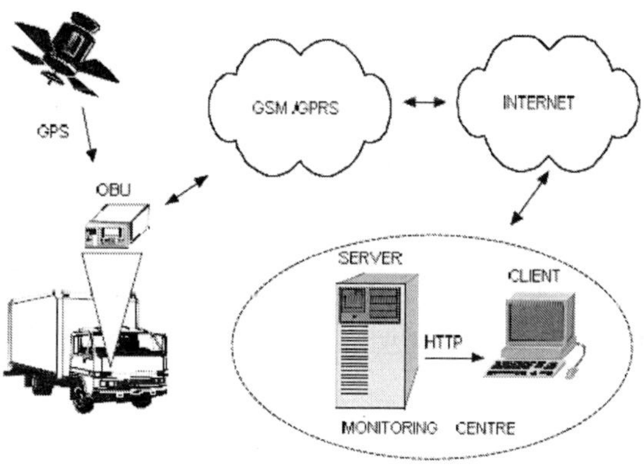

Figure 80: Communication setup (Vaculík et al., 2008)

5.8.2.3. Location information and mapping

ITS requires location information for detecting the position of a moving vehicle. Vehicles operate in a very limited or confined space (i.e., road). GNSS (Global Navigation Satellite System) is extensively used in ITS for locating position of the vehicle. However, GNSS does not provide accurate location data when the vehicle passes through obstructions like tunnels or while traveling through areas surrounded by high-rise buildings or mountains. Gyroscope and accelerometer are commonly installed for obtaining positions of the vehicles when it passes through the obstructions. In this manner, the location of the vehicle is detected. It is important to associate location data with map data as shown in Figure 81 to develop a dynamic map for indicating drivers with the ongoing driving route, best driving route, traffic congestion, etc. A map data is developed by integrating base map data with road network (link and node), road and traffic characteristics (signalized or un-signalized intersection, interchange, sidewalk, traffic constraint, etc.), static and dynamic information (landmark, traffic signs, time of day, weather forecast, road surface, traffic congestion, etc.). The development of map data will facilitate drivers to identify the best route for travel according to traffic congestion and time of day. It is very important to update map data after certain intervals in order to incorporate or identify road construction or road restriction or newly opened roads for providing the most appropriate information to the drivers.

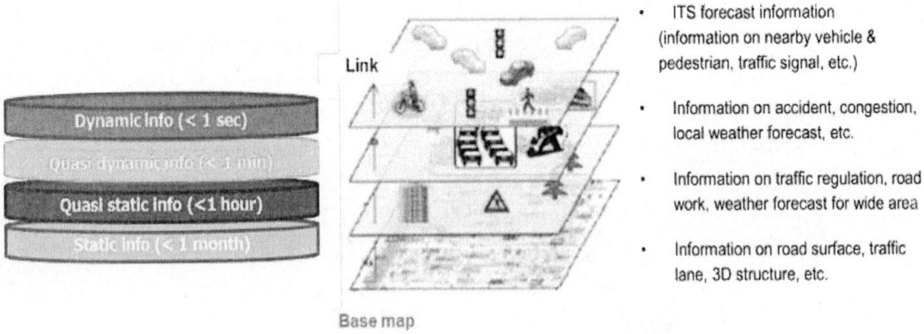

Figure 81: Concept of dynamic map (Maniko et al., 2016)

5.8.3. ITS applications

Road traffic accidents, traffic congestion, weather conditions, and other factors such as road blockage, visibility, and lighting issues, etc. can create problems in smooth traffic flow which may compromise safety of the road users. ITS can be effectively used to avoid these problems or minimize the impact of these problems on traffic flow and safety. The major ITS applications for improving traffic efficiency and road safety are:

- Advanced traveler information system is used for monitoring traffic flow for providing regular updates to the drivers regarding traffic congestion, road accidents, road blockage, road closures, different routes, and weather conditions.

- Electronic Toll Collection (ETC) can be deployed at tolls for automatic payment of toll amount for using toll roads.

- Emergency response management and emergency vehicle management provide facility of emergency notification and personal security to the road users. Road users can immediately notify the emergency service providers for assisting travelers encountering dangerous traffic situations like road traffic accidents either manually or automatically from the vehicle.

- Traffic operations at signalized intersections can be efficiently managed by automatically evaluating the traffic flow based on which signal timings can be appropriately controlled to reduce delay and increase the capacity of the intersection.

- The fleet size of public transportation can be optimized by analyzing data related to passenger demand, schedule, and vehicle location. This information can be integrated with GIS for maximum utilization of the public transport system.

- Intelligent vehicle highway system aims to improve safety by augmenting the ability of the drivers to control the vehicle while driving. This service assists drivers in avoiding collisions by providing advanced driver assistance and collision warning systems, adaptive cruise control, steering evasive systems, etc.

- The efficacy and safety of freight transport can be significantly improved by providing an electronic freight management system. The logistic operations can be improved through standardizing information exchange between warehouse operators, supply chain partners, and customer service. Further, ITS can play a major role in integration of different modes of transportation for efficient management of freight transport.

5.9. Public-Private Partnership in transportation

Road infrastructure is the backbone of the country's economic growth and development. A road network developed considering the aspects of mobility and accessibility highly facilitates transport of passengers and commodities which has a positive direct impact on a country's economic growth and overall development. India's economy, which is among the fastest expanding in the world, is predominantly dependent on the transportation industry. The demand for fast and efficient transport service is constantly growing every year. An underdeveloped road infrastructure cannot handle the ever-rising demand for fast and efficient transport service which negatively affects quality of life and slows down the economic progress. The government alone cannot improve and support the transport infrastructure. Thus, there is a need to form a Public-Private-Partnership (PPP) for collectively developing and expediting the process of improvement of road infrastructure.

5.9.1. Role of private sectors in transport development

Infrastructure development demands heavy investment with long remuneration periods. Initially, government used to develop the entire infrastructure of the country which included public utilities (telecommunication, water supply, electricity supply, sewage and solid waste management, gas supply, etc.) and public works (development of roads, railways, airports, and waterways) (Kadiyali, 2013). The government alone cannot handle the constant demand of improvement of the infrastructure for better services. Therefore, the government decided to privatize the development of infrastructure through PPP. PPP is a collaboration between public (government agencies at local, state, and central level) and private sectors formed for financing, constructing, operating, and maintaining the physical infrastructure of the country (International Civil Aviation Organization, 2022). There are multiple benefits that can be procured from privatization of the transportation infrastructure development. Inclusion of private sectors in infrastructure development will bring innovative ideas, additional resources, and cost-effective designs with better management skills which might expedite the project work. Thus, PPP can improve the quality, efficiency, and competitiveness of public services.

5.9.2. Different forms of public-private partnership

Public private partnership can be formed in different types as per the requirements and demands of public and private sectors as explained below:

- BOT: Build, Operate, and Transfer (BOT) is one of the most used forms of PPP which was successfully adopted for National Highway Projects in India. The public sector owns the project and the private sector builds, operates, and maintains the developed facility for a specified duration known as concession period. National Highway Projects in India were on a concession period of 20 years. The operator (private sector) transfers the facility in working condition to the public sector when the concession period ends. The operator recovers their investment by collecting user fees in the form of tolls (Kadiyali, 2013).
- BOOT: Build, Own, Operate, and Transfer (BOOT) is a type of PPP in which the facility is owned and developed by the private sector for the duration of the contract during which they recover the investment cost. The private sector handovers the developed facility to the government when the contract ends. BOOT structure is suitable when the government has a huge infrastructure financing gap and the private sector is prepared to take commercial risk and equity of the project for the term of the contract (Felsinger et al., 2008; Kadiyali, 2013).
- BOO: Build, Own, and Operate (BOO) is similar to BOOT. In BOO, there is no handover of the created facility to the public sector. This structure is suitable for projects with heavy investments and operating costs (Felsinger et al., 2008; Kadiyali, 2013).
- DBFOT: In Design, Build, Finance, Operate, and Transfer (DBFOT), the private sector designs, builds, finances, and operates the facility on a concession period. The private sector transfers the developed facility to the public sector when the concession period ends. It is similar to BOOT (Felsinger et al., 2008; Kadiyali, 2013).
- Special Purpose Vehicles: In this structure, the public and private sectors and other financial parties form a separate company and manage the entire project. The user fees collected are distributed among the stakeholders as per the agreement (Kadiyali, 2013).
- Annuity Project: In this structure, the concessionaire builds and operates the project using his/her own expenses and transfers the project to the government after the concession period. The concessionaire is not allowed to collect any user fees. Rather, the government pays a fixed amount at intervals to the concessionaire to cover his/her expenses (Kadiyali, 2013). Table 2 presents recent PPP projects conducted in India.

Table 2. PPP projects conducted in India (Department of Economic Affairs, 2019)

S. No.	Project name	Cost of project (in INR Crore)	Type of PPP	Concession period (in months)
1	Bus Terminal (Vadnagar) Project	10.53	BOT	36
2	Development of Offshore Multipurpose Cargo Berth at Mumbai Port in the State of Maharashtra	696.27	DBFOT	360
3	Machilipatnam Port Development	1860	BOOT	360

References and suggested readings

1. APSEd, 2022. Superelevation Concept and Formula Derivation with Practice Problem [WWW Document]. URL https://www.apsed.in/post/superelevation (accessed 6.21.22).
2. Chhabra, S., 2017. Overtaking Sight Distance Design- Highway Engineering.
3. Choudhary, P., Velaga, N.R., 2018. Performance Degradation During Sudden Hazardous Events: A Comparative Analysis of Use of a Phone and a Music Player During Driving. IEEE Trans. Intell. Transp. Syst. 20, 4055–4065. https://doi.org/10.1109/TITS.2018.2879968
4. Choudhary, P., Velaga, N.R., 2017. Modelling driver distraction effects due to mobile phone use on reaction time. Transp. Res. Part C Emerg. Technol. 77, 351–365. https://doi.org/10.1016/j.trc.2017.02.007
5. CRRI, 2017. Indian- Highway Capacity Manual.
6. Department of Economic Affairs, 2019. List of Infrastructure Project [WWW Document]. Gov. India. URL https://www.pppinindia.gov.in/infrastructureindia/project list?id=1&searchType=Government Infrastructure Projects (PPP) (accessed 6.12.22).
7. ESE notes, 2022. Transition curve and length of transition curve [WWW Document]. URL https://esenotes.com/transition-curves-length-of-transition-curve/ (accessed 6.21.22).
8. Essex, D., 2022. Logistics [WWW Document]. URL https://www.techtarget.com/searcherp/definition/logistics (accessed 6.12.22).
9. Felsinger, K., Miranda, J., Skilling, H., Booth, K., Areneta, E., Birken, M.A., Pedersen, S., Edwards, S., Woodward, I., Herrera, V., Canzon, M., Sutarez, A., Mannapbekov, N., 2008. Public-Private Partnership Handbook. Asian Dev. Bank 100.
10. Indian Railways, 2017. Reform, Perform and Transform.
11. Indian Railways Civil Engineering Portal, 2022. Indian Railways [WWW Document]. URL https://ircep.gov.in/AboutUs.html#main-content (accessed 6.21.22).
12. International Civil Aviation Organization, 2022. Public-Private Partnership (PPP) [WWW Document]. URL https://www.icao.int/sustainability/pages/im-ppp.aspx#:~:text=The advantage of a PPP, the competitiveness of public services.

(accessed 6.21.22).
13. Kadiyali, L.R., 2013. Traffic engineering and transport planning.
14. Khanna, S.K., Justo, C.E.G., Veeraragavan, A., 2014. Highway Engineering.
15. Khanna, S.K., Justo, C.E.G., Veeraragavan, A., 2013. Highway materials and pavement testing.
16. Mahajan, K., Velaga, N.R., 2021. Sleep-deprived car-following: Indicators of rear-end crash potential. Accid. Anal. Prev. 156, 106123. https://doi.org/10.1016/j.aap.2021.106123
17. Maniko, H., Kamijo, S., Shin, C.-H., Chung, E., 2016. Intelligent Transport Systems (ITS) Introduction Guide.
18. Mathew, T. V., 2019. Cross Sectional Elements [WWW Document]. URL https://www.civil.iitb.ac.in/tvm/nptel/302_CroSecEle/web/web.html#x1-40002.1 (accessed 6.21.22).
19. Mathew, T. V., 2006. Transportation Engineering - I.
20. Ministry of Civil Aviation, 2019. Vision 2040 for the Civil Aviation Industry in India.
21. Ministry of Ports Shipping and Waterways, 2021. Maritime India Vision 2030 300.
22. Ministry of Ports Shipping and Waterways Government of India, 2018. Inland Waterways Authority of India [WWW Document]. URL https://iwai.nic.in/about-us (accessed 6.21.22).
23. Ministry of Road Transport & Highways, 2020. Road Accidents in India - 2019.
24. Ministry of Road Transport and Highways, 2017. Bharatmala Pariyojana Phase-1.
25. Pawar, N.M., Gore, N., Arkatkar, S., 2022a. Examining Crossing Conflicts by Vehicle Type at Unsignalized T-Intersections Using Accepted Gaps: A Perspective from Emerging Countries. J. Transp. Eng. Part A Syst. 148, 1–15. https://doi.org/10.1061/jtepbs.0000665
26. Pawar, N.M., Velaga, N.R., 2022. Analyzing the impact of time pressure on drivers' safety by assessing gap-acceptance behavior at un-signalized intersections. Saf. Sci. 147, 105582. https://doi.org/10.1016/j.ssci.2021.105582
27. Pawar, N.M., Velaga, N.R., 2021. Effect of time pressure on steering control of the drivers in a car-following situation. Transp. Res. Part F Traffic Psychol. Behav. 80, 218–236. https://doi.org/10.1016/j.trf.2021.04.007
28. Pawar, N.M., Velaga, N.R., 2020. Modelling the influence of time pressure on reaction time of drivers. Transp. Res. Part F Traffic Psychol. Behav. 72, 1–22. https://doi.org/10.1016/j.trf.2020.04.017
29. Pawar, N.M., Velaga, N.R., Mishra, S., 2022b. Impact of time pressure on acceleration behavior and crossing decision at the onset of yellow signal. Transp. Res. Part F Psychol. Behav. 87, 1–18. https://doi.org/10.1016/j.trf.2022.03.014
30. Saket, S., 2020. Economic Survey 2019-20. Inst. Policy Res. Stud. https://doi.org/10.1016/0306-4549(82)90023-8
31. Soehodho, S., 2009. Road accidents in Indonesia. IATSS Res. 33, 122–124.
32. Vaculík, M., Franeková, M., Vestenický, P., Vestenick, M., 2008. On-Board Unit and its Possibilities of Communications on Safety and Security Principles. Adv. Electr. Electron. Eng. ISSN 1336-1376 7, 235–238.
33. World Health Organization, 2018. Global status report on road safety 2018.

34. Yadav, A.K., Velaga, N.R., 2021. Modelling brake transition time of young alcohol-impaired drivers using hazard-based duration models. Accid. Anal. Prev. 157, 106169. https://doi.org/10.1016/j.aap.2021.106169

Dynamic QR Code for Further Reading

1. Road accident statistics of India: https://morth.nic.in/sites/default/files/RA_Uploading.pdf	
2. Indo-HCM Manual: https://crridom.gov.in/sites/default/files/files/Indo-HCM%20Snippets.pdf	
3. Public-Private Partnership: https://www.adb.org/sites/default/files/institutional-document/31484/public-private-partnership.pdf	

4. Introduction to ITS: https://www.jsce-int.org/system/files/ITS_Introduction_Guide_2.pdf	

6 Practical Activities

Unit specifics

Through this unit we have discussed the following aspects

- *Bridge model construction*
- *Preparing building estimations*
- *Structural stability*
- *Application of Geographic information system (GIS)*
- *Dam model construction*
- *Basics of technical writing*
- *Application of python for data handling*

The practical applications of the topics are discussed for generating further curiosity and creativity as well as improving problem solving capacity.

Besides giving a large number of multiple-choice questions as well as questions of short and long answer types marked in two categories following lower and higher order of Bloom's taxonomy, assignments through a number of numerical problems, a list of references and suggested readings are given in the unit so that one can go through them for practice. It is important to note that for getting more information on various topics of interest some QR codes have been provided in different sections which can be scanned for relevant supportive knowledge.

After the related practical, based on the content, there is a "Know More" section. This section has been carefully designed so that the supplementary information provided in this part becomes beneficial for the users of the book. This section mainly highlights the initial activity, examples of some interesting facts, analogy, history of the development of the subject focusing the salient observations and finding, timelines starting from the development of the concerned topics up to the recent time, applications of the subject matter for our day-to-day real life or/and industrial applications on variety of aspects, case study related to environmental, sustainability, social and ethical issues whichever applicable, and finally inquisitiveness and curiosity topics of the unit.

Rationale

This unit on practical activities will allow students to develop basic skills about the elementary practical applications of civil engineering. It covers various fun and practical tasks related to computational and manual applications in civil engineering. This unit covers manual tasks like bridge model construction with popsicles, estimation of a house plan, building frames to explore stability under vertical loading, and simple dam construction. The materials prescribed for the tasks are very light and easily available. It also contains computational tasks of application of geographic information system (GIS) and programming. Computational proficiency is very important in both academic and professional fields to handle large data and complex computations. This chapter also discusses about technical writing which is one of the important parts of any engineering and scientific study. The tasks in the chapter will help the students to develop a preliminary mindset of application-based work that are relevant in the field of civil engineering

Pre-requisites

Mathematics: Coordinate Systems (Class XII)

Physics: Mechanics (Class XII)

Unit outcomes

List of outcomes of this unit is as follows:

U1-O1: Build a popsicle bridge

U1-O2: Understand basics of estimation.

U1-O3: Explore structure stability under vertical loading.

U1-O4: Preliminary idea about application Geographic Information System (GIS) application in geology.

U1-O5: Explore ideas of dam construction.

U1-O6: Understand the basics of technical writing.

U1-O7: Preliminary introduction to programming for handling large data sets.

Unit-1 Outcomes	***EXPECTED MAPPING WITH COURSE OUTCOMES*** *(1- Weak Correlation; 2- Medium correlation; 3- Strong Correlation)*					
	CO -1	***CO -2***	***CO -3***	***CO -4***	***CO -5***	***CO -6***
U1-O1	*3*	*3*	*3*	-	*3*	*1*
U1-O2	*1*	*1*	*2*	*2*	*1*	-
U1-O3	*2*	*1*	*3*	*1*	*2*	*1*
U1-O4	-	-	*3*	*1*	*2*	*2*
U1-O5	*3*	*3*	*3*	-	*3*	*1*

6.1. Bridge model construction

The objective of this activity is to design a bridge with maximum load capacity and is aesthetically pleasing. The challenge is to optimize the material usage since the available resources will be only 200 popsicle sticks and glue. The bridge model is expected to look like the picture as shown in Figure 82. Teams are supposed to test the bridge load capacity in presence of other teams and present the findings to the class.

Resources:

Each group of students will be provided with the following materials:

1. 200 popsicle sticks, a glue gun
2. 10 Kg weights or equivalent weights using a bag of sand.

Figure 82: Popsicle Bridge (Source: https://tryengineering.org/)

Time Needed: Two 45-minute sessions.

Instructions:

- Construct a team of 4-5 students and gather the required materials.
- Understand the constraints of materials and along with the objective of maximum load bearing capacity.
- The students shall be encouraged to design an aesthetically pleasing structure.
- The minimum span of the bridge shall be 35 centimeters.
- After the instructions, students shall discuss among team members and design a layout on the A4 paper.
- Students shall discuss the plan with the instructor and may need time to rethink and redesign the layout if needed.
- After completion of layout, the students shall execute and complete the construction of the bridge.
- After completion of the bridge, the constructed bridges shall be placed one meter above the ground (can be placed in between two chairs).
- The bridge now shall be checked for maximum load capacity. The load shall be increased gradually, and the structure shall bear the maximum load for at least 1 minute.
- Each bridge shall be peer evaluated for aesthetics. All the students will rate the other team member's bridge on a scale of 1-5. (1: not at all appealing; 2: not appealing; 3: neutral/average; 4: somewhat appealing; 5: very appealing).
- Further, the students can share their learning and the possible improvements they would make if they could design the bridge again.
- A discussion should be initiated regarding the tradeoff between functionality, safety, and aesthetics for any structural building.

6.2. House plan and model

To make an estimated bill of the quantities required for the above-ground construction of a 1 Bedroom-Hall-Kitchen (BHK) flat (Figure 83).

Resources: A4 Sheets, Calculator, Pencil, and Eraser.

Time Needed: 45-minutes.

Instructions:

- Divide students into teams of three.
- Students shall draw the front and top view of the bedroom, hall, and kitchen.

- Prepare the list of the structures in all three substructures of 1 BHK house.
- Break down the structures into the element and identify the required building materials.
- Calculate the approximate quantity of material required in the structures based on their dimensions.
- Estimate the cost of the building materials and find out the total Bill of Quantity (BOQ).
- Compare the BOQ with the other teams and identify if any structures were missed

Learning Objectives:

- To be able to visualize a building structure and draw it on paper.
- To develop an understanding of the costs of the building construction materials.
- To familiarize with the billing process of the construction sites.

Sample BOQ:

Item	Description	Qty	Unit	Rate	Total Amount
				(SR)	(SR)

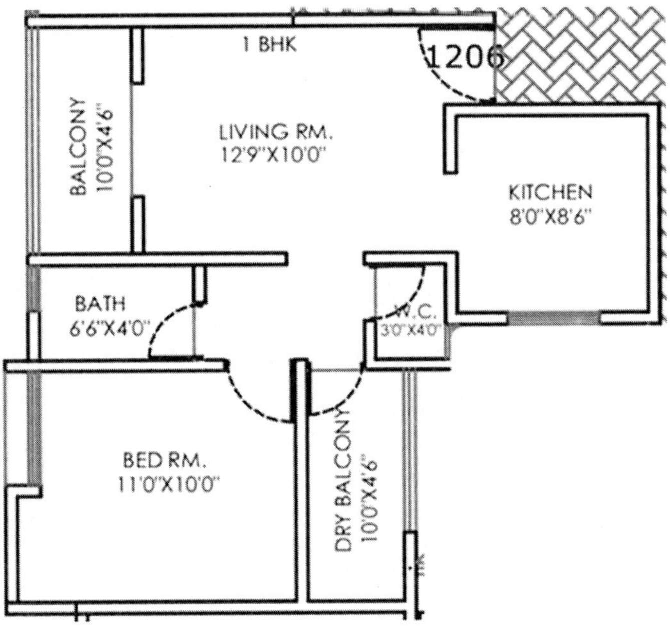

Figure 83: Sample Sketch of a House

6.3. Stability of structures

Source: pbslearningmedia.org

In this activity students will learn about the stability of different shapes under vertical load. The structure showing the minimum deflection will be the most stable.

Resources:

1. 7 drinking straws
2. 14 paper clips

Time Needed: 45-minute session.

Instructions:

- Make a group of 2-3 students and provide the material per group.
- Students shall make a prediction about the stability of the structure.
- Students shall build a triangle and a square frame from the straws and paper clips.
- The paper clips can be further used to connect the straws in the required shapes.
- Each shape shall be tested by applying the downward force on the top.
- Discussion on the stability of the shape and their reasoning shall be performed after the testing.

Further Discussion:

- Is further reinforcement of the structure possible by adding additional straws and paper clips? If yes, discuss the possibilities with minimum additional resources.
- Can you build a structure using the most stable shapes and how much weight will that structure be able to hold?

6.4. Geology/Soil:

In this section, students will get familiar with the geological and soil-related data using Q-GIS (Geographic Information System) for preliminary understanding with a simple hand on practice. It is open-source software for processing and visualizing GIS datasets.

Before going into detail about GIS processing, you should know about shapefiles, raster, and vector files to better understand geological datasets. A shapefile is a storage of vector data in which shape, location, and characteristics of geographic features are stored in the .shp extension with .shx, .dbx, and .prj auxiliary files. GIS datasets are stored in vector and raster data structures. Vector datasets generally store geographic information in lines, arcs, and polygons, while raster datasets are in a fixed pixel or gridded format.

There are lots of software and packages to process geology or soil-related GIS datasets. In this section, we will be using Q-GIS for mapping and visualizing the global soil map and clip it for India. It can be downloaded and installed from the QGIS website (https://www.qgis.org/en/).

Soil map in the form of shapefile for the world has been downloaded from http://www.fao.org/soils-portal/soil-survey/soil-maps-and-databases/faounesco-soil-map-of-the-world/en/ website.

The downloaded shapefile (.shp format) is added in Q-GIS as a vector layer. A screenshot (Figure 84) has been added here to visualize the digital soil map of the world (DSMW).

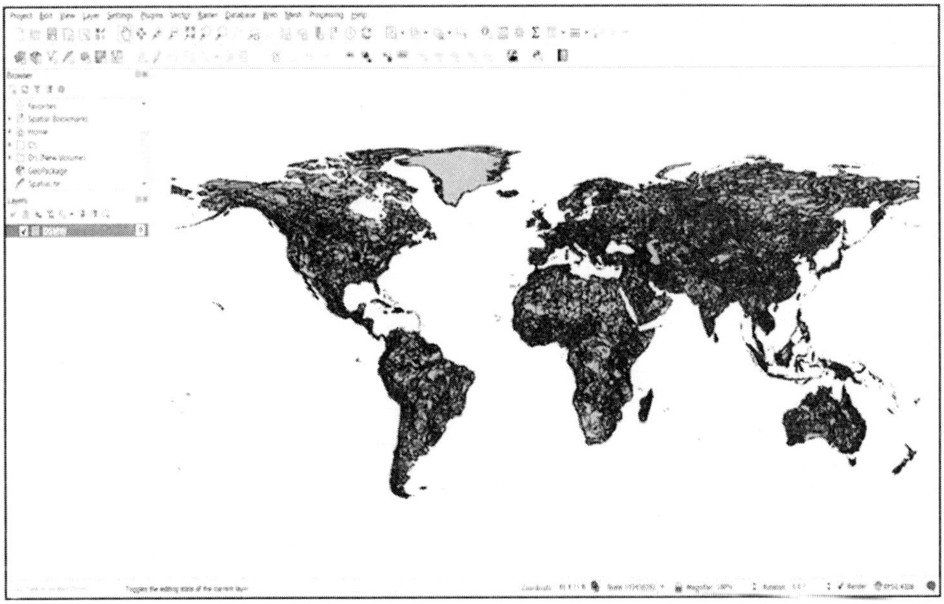

Figure 84: A screenshot of the Q-GIS showing a digital soil map of the world

DSMW is a shapefile containing soil type information for the countries of the globe. Our task is to clip it for India and observe the spatial variation of soil type throughout the country. The shapefile's attribute table is shown in Figure 85, which contains information such as country names, soil type codes, polygon area, etc.

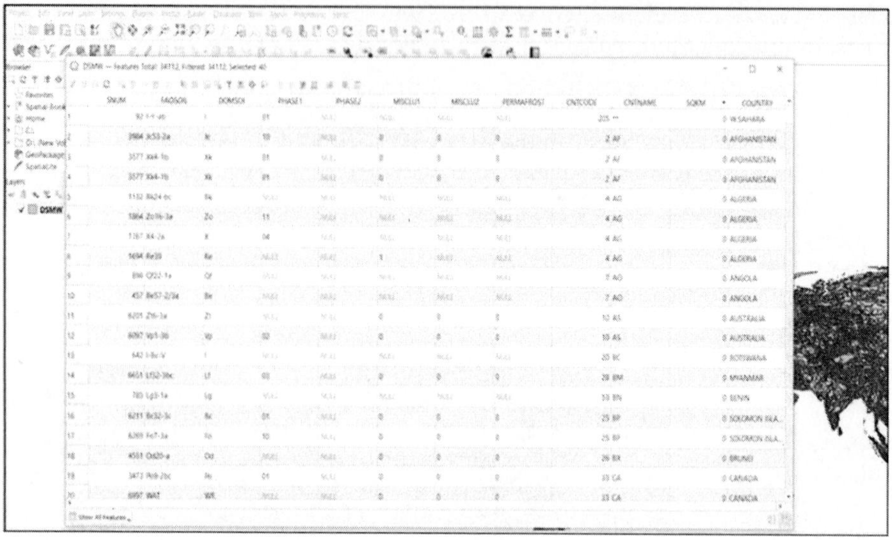

Figure 85: A snapshot of the attribute table of the DSMW

To subset DSMW for India, the attribute table rows reflecting country name as India are selected and exported as a new shapefile which will contain the soil map of India (Figure 86).

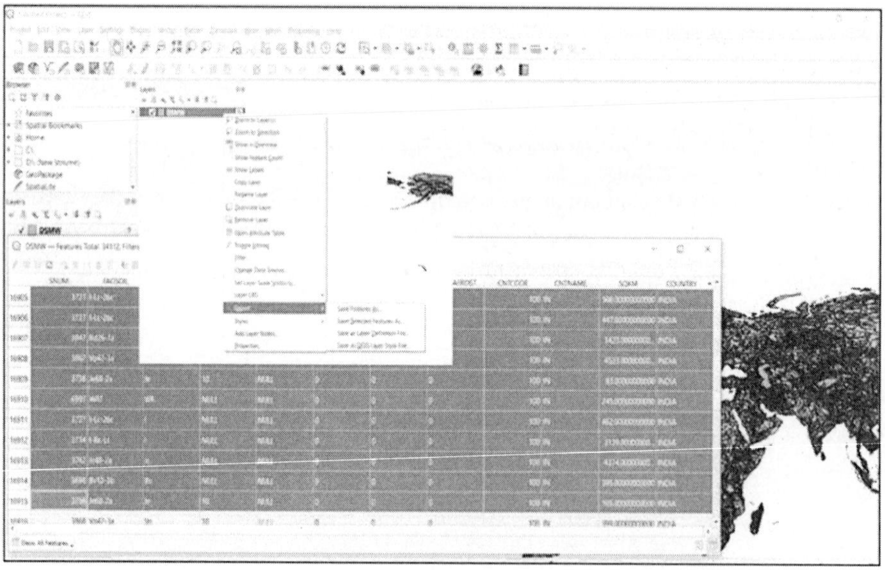

Figure 86: A snapshot of exporting soil characteristics for India

The clipped soil map, i.e., DSWI (Digital Soil Map of India), is now added to Q-GIS for visualizing the spatial variation of soil types in the country. The readme file downloaded with the DSMW folder details the abbreviations used for soil type codes. The final soil map for India is shown in Figure 87, with full names of the soil types.

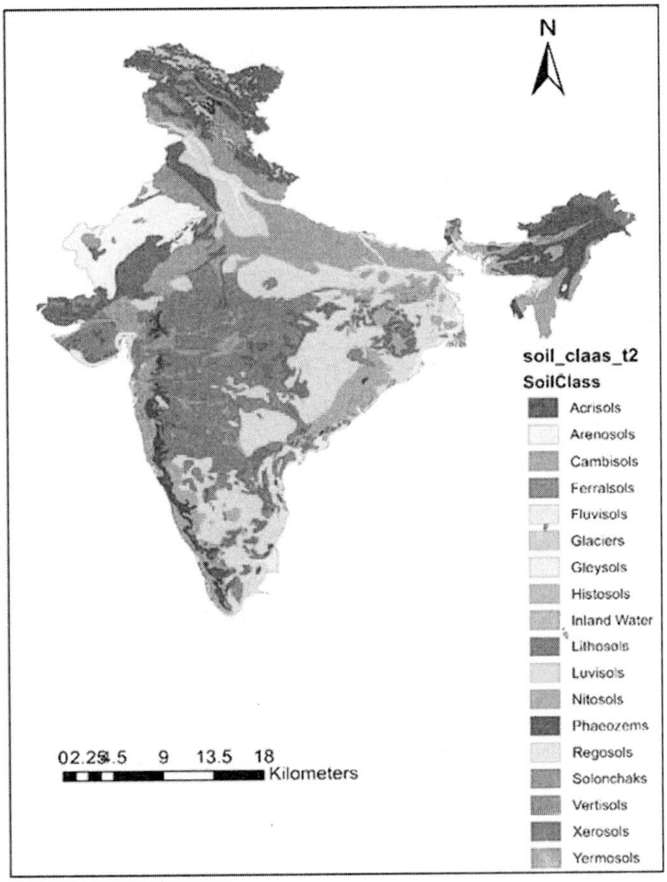

Figure 87: Digital soil map of India containing soil types (Source: FAO-UNESCO Soil Map of the World)

Can you answer these questions on soil mapping using GIS?

- What is the highest and lowest spread area of soil types in India?
- What is the percentage spread area corresponding to each soil type in each state?
- Can you plot the percentage spread area corresponding to each soil type over India?
- Can you plot the percentage spread area corresponding to each soil type for all states?

6.5. Dam Construction

Source: *https://tryengineering.org/*

The objective of this study is to understand the functioning and design of dam construction by creating a system of damming water in a trough. The system must have the capacity to hold the water and release the water in a controlled manner.

Resources:

Every group must have the following materials:

1. Long plastic box (Figure 88).
2. Gravel or sand
3. Water
4. Cardboard
5. PVC pipes
6. Tape
7. Foil

Figure 88: Long plastic planter box (Source: https://tryengineering.org/)

Time Required: 90- 120 minutes.

Instructions:

- Each group shall consist of 3-4 students.
- Brief discussion on the dam construction challenges and the design criteria and process shall be conducted.
- Plot the sketch of the design and further remodification shall be discussed with the instructor so that each system can hold up to 5 liters of water.
- Thereafter each group shall work on building their designs and further testing of the dam systems shall be conducted to evaluate the performance.

Evaluation Criteria:

The following evaluation criteria shall be checked to rate the performance of the system:

- Did the dam hold the water back?

○10 points (No water escaped)

○5 points (Some water escaped)

○0 points (All water escaped)

- Were you able to release water and then stop it again?

○10 points: yes

○0 points: no

- Involvement of individual group member and the teamwork?

○10 points: yes

○0 points: no

- Discussion and question answer session.

6.6. Technical writing

As a civil engineering student, you have been tasked to write a term paper on structural health monitoring or the hydrologic cycle. How do you proceed to write with technical clarity? Most of the time, you may have certain ideas based on the topic, but it takes considerable effort to connect your ideas and express your knowledge in a well-written document. Will you start from the individual terms such as evaporation, interception, transpiration, precipitation? Or will you connect the points into a story from the beginning? Before you write, you should first develop a basic structure of your document. And then decide what level of information you would need to put in the paper. What format would you have to use? The technique of documenting information or procedures in the fields of industry, academia, and research is known as technical writing. Also referred to as academic writing if you are writing in an academic environment.

Avoid all plagiarism when writing a technical document. Have you ever copied your friend's assignment or any kind of academic documents, changed a few words, and submitted it as yours? That is a textbook case of plagiarism. Plagiarism can be defined as "the practice of taking someone else's work or ideas and passing them off as one's own." One can copy someone's idea in their document by acknowledging or citing their name. Not only the idea but also all other pieces of work such as images, video, or data etc. The consequences of plagiarism can be disastrous as universities can revoke the degree that has been granted to you or in some Countries it is punishable by law as well.

While writing a technical report or document, one of the most important things is to identify the target reader and identify their level of knowledge of that subject. You have to understand your readers and present the idea based on their prior understanding. For engineers, you can write the high-level description without much detailed introduction, whereas for school students you should provide more detailed information in the introductory section. A well-written document will explain things in a way that it anticipates and answers all the questions that a reader might have.

Once the target reader has been identified, you have to organize the knowledge into a document by following the correct order of presenting your ideas. If you want your writing to be clear, the most impactful thing you can do is to master this order. In order to remember new material, the human brain constantly organizes data into distinct categories. Therefore, it is recommended that the document be written in a structured fashion, with related sections written separately. There are two strategies for presenting our ideas in a well-written document - top-down approach or bottom-up approach. The top-down approach starts with a summary word or main topic and goes a step down by logically categorizing it into several ideas and further breaking down these individual ideas into several subsections as shown in the Figure 89. Consider an example topic called Engineering. One of the logical categorizations would be the different branches we could divide Engineering into - Civil Engineering, Computer Science Engineering, Electrical Engineering and Mechanical Engineering. Again, Civil Engineering can be categorized as Water Resource Engineering, Transportation Engineering, Structural Engineering etc. Similarly, Computer Science can be further categorized as Software Engineering, Network and Security, Machine Learning etc.

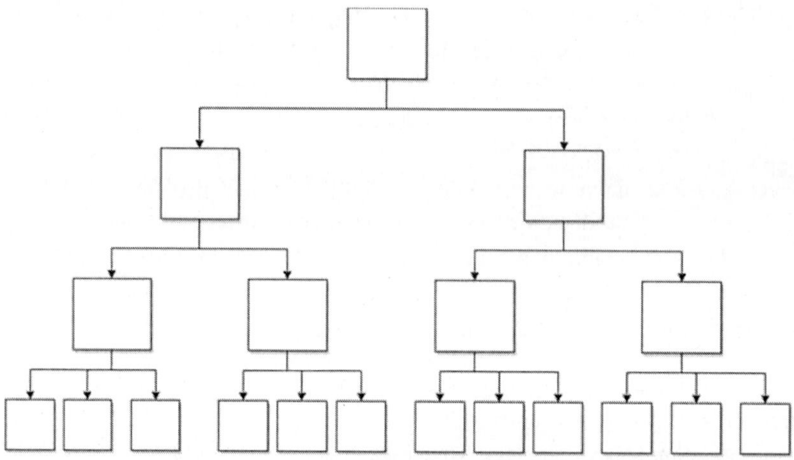

Figure 89: Writing usually calls for a pyramid of ideas centered on a particular topic (Source: Minto, B., The Pyramid Principle, 2009)

Whereas in the bottom-up approach the logically connected sentences or ideas form a paragraph, then logically connected paragraphs form a subsection, and logically related subsections are grouped into a section and so on. It is always better for a beginner to start with a top-down approach since it easily gives the structure of a pyramid. The pyramid's concepts at every level should always be abstracts of the information grouped beneath them. There should always be the same type of concept in each group.

When writing is structured in the form of the pyramid, it becomes easier for the human brain to parse the ideas. All of these summaries or sections should be connected to one another in the pyramid's top to bottom direction. This technique is known as vertical relationship. Similar to this, in the horizontal relationship, every item in a pyramid level should be connected to every other item in the same level of abstraction. The type of concepts you require in each grouping can be determined by knowing the vertical relationship. Knowing the horizontal relationship allows you to determine whether the thoughts you combine are of a similar kind. You can move on to the next stage if each level of the pyramid meets these two requirements.

For each technical document, there will be a different format. The primary objective of the technical document is to convey your technical knowledge or information to the readers. The introduction of a technical document typically comes first. The reader is informed about what they already know about the topic or what they should be expected to know in the introduction. These three components of a story must always be present in the introduction: the problem, the answer, and the circumstance. The introduction's length should reflect both the needs of the reader and the requirements of the topic.

Pay attention to how the pyramid's tiers change from one another. After writing the opening and body of your document, you must pause sometimes to inform the reader of your past and future destinations. Summarizing gives an additional strength to the technical documents. Suppose each chapter is extremely long, then summarizing the entire chapter using crisp language helps the reader to understand the whole idea of that chapter.

Within the pyramid, each grouping must always contain logically ordered concepts. In other words, there must be a reason why the second thought is placed second and not first or third. There are only four reasonable ways to arrange a list of ideas:

- Deductively
- Chronologically
- Structurally
- Comparatively

In deductive grouping, the major premises are presented first, followed by the minor premises and a conclusion. In the remaining three groups, however, ordering can be determined by preference. Ideas collected in writing are never the result of chance. They are always selected based on the author's logical analysis. Chronological grouping is among the simplest classifications as it follows the time order. The steps a person must take to accomplish a specific effect are laid down in a time-ordered grouping, together with the sequence in which he must do them. Structural order is the arrangement that corresponds to what you perceive after seeing something, whether through a diagram, a map, a drawing, or a photograph. You can envision a process or an object that is actual or conceptual. In ranking a group or comparative group, the arrangement of ideas depends entirely on how crucial each item is. This is the hierarchy you place on a bunch of things you've grouped together because you've determined that they all share a certain quality.

Some important things to keep in mind when writing technical documents are that they should always be neutral and devoid of overtly expressive language. Avoid using redundant words and concepts as the reader won't be interested in reading a document that uses unnecessary words. Try to explain the solution in a technical document rather than using exclamatory phrases and questions. Proper use of grammar and punctuation is another strength of a technical document. It makes the document more efficient and organized. Always write the information in your own language and cite the document properly to avoid plagiarism by using an open-source tool such as Zotero (Source: https://www.zotero.org/)

Adopt an optimistic tone in your writing. Instead of dazzling your readers with flowery words, utilize clear and simple words to express your idea because technical documents by nature are tough to comprehend already. Additionally, avoid contractions and unacademic abbreviations in the document. In essence, you must translate complex technical ideas into a language that your readers can understand.

6.7. Computational Methods:

Civil Engineering projects produce voluminous amounts of data, and we require sophisticated tools to analyze them. Writing computer programs using programming languages is a powerful skill that civil engineers employ to accomplish custom tasks related to modeling, analysis, interpretation, and visualization. Some of the most popular languages used in civil engineering are Python, C++, FORTRAN, R, etc.

Python is a free and powerful object-oriented programming language that works on all popular operating systems (Windows, MacOS, and Unix). Python is a popular platform for a wide number of applications, including web development, automation, desktop GUI applications, software and game development, data analysis and visualization, and machine learning. The relatively simple syntax of the language allows users to read, understand, and write code succinctly, allowing for rapid iteration and fast development cycles. For a beginner in programming, Python is one of the best choices for handling large data, modeling, prototype creation, and visualization. Python learning resources are widely available on the internet and this section only provides an example of a civil engineering application.

This section provides an example from Water Resources Engineering, where we use Python to perform a simple analysis using a long time-series of streamflow data.

6.8.1. Download hydrological data:

For this task, we require a long river streamflow dataset. Monthly mean streamflow time series can be downloaded from the GRDC (Global Runoff Data Center) website (https://portal.grdc.bafg.de/applications/). GRDC is a World Meteorological Organization (WMO) accredited data center and its two-hundred-year-old archive supports global long-term hydrological investigations.

First, we download the data file corresponding to the "Pandu" station (Filename: 2851300_Q_Month.txt) in the state of Assam, India, the data is available in text format and contains basic information about the station and the data available. A snapshot of the data file is provided in Figure 90 for quick reference to the original data downloaded.

```
# Title:                    GRDC STATION DATA FILE
#                           --------------
# Format:                   DOS-ASCII
# Field delimiter:          ;
# missing values are indicated by -999.000
#
# file generation date:  2022-06-10
#
# GRDC-No.:                 2851300
# River:                    BRAHMAPUTRA RIVER
# Station:                  PANDU
# Country:                  --
# Latitude (DD):          26.13
# Longitude (DD):         91.7
# Catchment area (km²):
# Altitude (m ASL):             49.0
# Next downstream station:        -
# Remarks:
# Owner of original data: Initial dataset collected in the framework of the First
GARP Global Experiment (FGGE)
#************************************************************
#
# Data Set Content:         MEAN MONTHLY DISCHARGE (MQ)
#                           --------------------
# Unit of measure:                    m³/s
# Time series:              1956 - 1979
# No. of years:             24
# Last update:              2018-05-25
#
# Table Header:
#       YYYY-MM-DD - Date (DD=00)
#       hh:mm      - Time
#       Original   - original (provided) data
#       Calculated - GRDC calculated from daily data
#       Flag       - percentage of valid values used for calculation from daily
data
#************************************************************
#
# Data lines: 287
# DATA
YYYY-MM-DD;hh:mm; Original; Calculated; Flag
1956-01-01;--:--;   3970.000;   -999.000;    0
1956-02-01;--:--;   -999.000;   -999.000;    0
```

Figure 90: A snapshot of the data file downloaded from the GRDC website.

Task: The task is to write a Python script that takes this raw text data, cleans it, processes it, and finally visualizes it. The original data is from 1956-1979, but we will only perform the analysis from 1971, which will allow us to learn Python operations of sub-setting data. convert the given text format data into a data frame and plot the time-series after the year 1971. Thereafter, we must find the number of times the monthly mean discharge crossed 10,000 cumecs and find out the missing number of monthly mean values from 1971 to 1979.

6.8.2. Design of the Python script:

Step 1: Import libraries required for the task. Here, Pandas is used to convert the text file into a data frame and analyze the data. While Matplotlib will be used for plotting the data.

```
import pandas as pd
import matplotlib.pyplot as plt
import matplotlib.dates as mdates
plt.style.use('fivethirtyeight')
```

Step 2: Read the raw data using pandas and assign it to the variable "data". The parameter "comment" will instruct the program to exclude the lines that start with "#" from the analysis. Since the values themselves are separated by ";," the parameter "sep" allows sorting the continuous text into different columns. At the same time, "na_values" is used to specify the default NaN value in the dataset. The values and symbols assigned to each parameter are based on the raw data.

```
data = pd.read_csv(r'F:\projects\aicte\2022-06-10_06-27\2851300_Q_Month.txt',comment='#', sep=';|;', na_values=('-999'))
```

The output of this command is shown in Figure 91 which now shows 287 columns and 5 rows stored in the variable "data" for further analysis.

	YYYY-MM-DD	hh:mm	Original	Calculated	Flag
0	1956-01-01	--:--	3970.0	NaN	0
1	1956-02-01	--:--	NaN	NaN	0
2	1956-03-01	--:--	5670.0	NaN	0
3	1956-04-01	--:--	7650.0	NaN	0
4	1956-05-01	--:--	25400.0	NaN	0
...	...	...	...	...	...
282	1979-07-01	--:--	37967.0	NaN	0
283	1979-08-01	--:--	49210.0	NaN	0
284	1979-09-01	--:--	33619.0	NaN	0
285	1979-10-01	--:--	30736.0	NaN	0
286	1979-11-01	--:--	1033.0	NaN	0

287 rows × 5 columns

Figure 91: Output of step 2

Step 3: In order to make the analysis easier, we name the relevant columns.

```
data = data[['YYYY-MM-DD','Original']]
data.rename(columns={'YYYY-MM-DD':'Date','Original':'Streamflow (cumecs)'}, inplace=True)
data
```

Output: Figure 92

	Date	Streamflow (cumecs)
0	1956-01-01	3970.0
1	1956-02-01	NaN
2	1956-03-01	5670.0
3	1956-04-01	7650.0
4	1956-05-01	25400.0
...	...	...
282	1979-07-01	37967.0
283	1979-08-01	49210.0
284	1979-09-01	33619.0
285	1979-10-01	30736.0
286	1979-11-01	1033.0

287 rows × 2 columns

Figure 92: Output of step 3

Step 4: Finding the data type of all the columns in the data frame. Figure 12 shows that Date is an object, and Streamflow is a float64 type.

```
data.dtypes
```

Output: Figure 93

```
Date                    object
Streamflow (cumecs)    float64
dtype: object
```

Figure 93: Output of step 4

Step 5: Convert the data type of the "Date" column to datetime format of Python. This makes the dataset suitable for time series analysis.

```
data['Date'] = pd.to_datetime(data['Date'])
data.dtypes
```

Output: Figure 94

```
Date                          datetime64[ns]
Streamflow (cumecs)                  float64
dtype: object
```

Figure 94: Output of step 5

Step 6: Subset all data after the year 1971 and store it in a new variable (data_1971)

```
data_1971 = data[(data['Date'] > '1971-1-1')]
```

Step 7: Plotting the streamflow time-series

```
years = mdates.YearLocator()
years_fmt = mdates.DateFormatter('%Y')
ig,ax = plt.subplots()
ax.plot(data_1971['Date'],data_1971['Streamflow (cumecs)'], color ='blue')
ax.xaxis.set_major_locator(years)
ax.xaxis.set_major_formatter(years_fmt)
plt.xlabel("Year", weight='bold')
plt.ylabel("Streamflow (cumecs)", weight='bold')
plt.legend()
```

Output: Figure 95

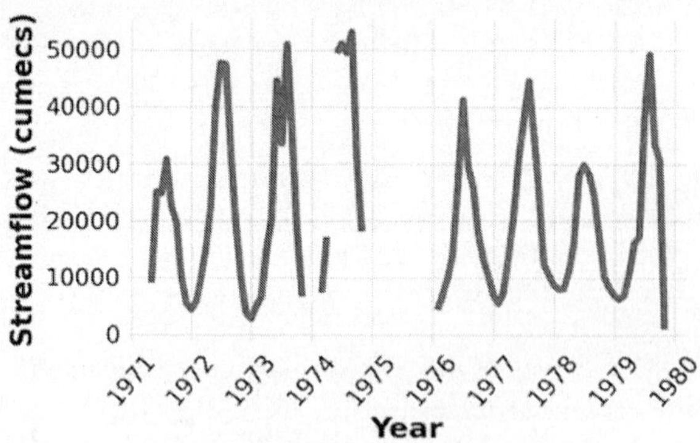

Figure 95: Plot of the stream flow time series

Step 8: How many times has the streamflow crossed 10000 cumecs? We can answer this using the following code.

```
(data_1971['Streamflow (cumecs)'] > [illegible]).sum()
```

Output:

```
59
```

Step 9: Finding the number of NaN values in the Streamflow data

```
data_1971.isna().sum()
```

Output: Figure 96

```
Date                   0
Streamflow (cumecs)   21
dtype: int64
```

Figure 96: Output of step 9

As you can see, this python code allows extreme flexibility in answering questions of our interest. The best way to learn any programming language is by implementing real-world projects and analyses. Can you answer the following questions using Python?

Q1. Compute the average annual flow for the years 1971 to 1979.

Answer:

1971: 18469.38 cumec

1972: 21521.17 cumec

1973: 21657.36 cumec

1974: 34648.38 cumec

1975: NaN

1976: 18207.36 cumec

1977: 20825.33 cumec

1978: 16554.58 cumec

1979: 19707.09 cumec

Q2. Are high and low streamflow observations confined to any specific month(s)? If so, which months are they?

Q3. Find the maximum and minimum flow and their corresponding dates over the entire time series.

Answer: Maximum 56500 cumecs on 01-07-1959 and minimum is 1033 cumecs on 01-11-1979.

Q4. Plot the time series of all the July months from 1971.

Q5. Find the number of times there is a dip in the streamflow magnitude compared to its previous day flow from 1971.

6.8. Entrepreneurial opportunities in civil engineering

The process of planning, starting, and running a business is called entrepreneurship. An engineer is expected to find innovative and scientific solutions to problems in society. These ideas and solutions can be implemented to start and run a business. The creative mindset and problem-solving attitude of engineers are vital for success in entrepreneurship. However, an engineer must develop management and leadership qualities to successfully run a business.

Entrepreneurship in the field of civil engineering requires an understanding of real-world engineering problems, in-depth knowledge of relevant subjects, and the skill with the tools necessary to accomplish the task. Entrepreneurship in civil engineering mainly deals with the design and construction of structures. However, there are other aspects of civil engineering where the possibility of entrepreneurship exists. Some of the contemporary civil engineering problems where entrepreneurial opportunities can be explored are listed as follows:

- Innovative design and construction of low-cost housing
- Innovative solutions for smart and green buildings
- New innovative construction and building materials
- Innovative waste management systems
- Smart traffic management systems
- Develop software for civil engineering applications.
- Data analytics related to the built and natural environment.

There are also popular entrepreneurial practices such as running structural design and construction companies, hydraulic design companies, soil testing, and ground improvement companies, survey firms, civil engineering consultancy firms, etc. Entrepreneurship directly impacts the economic growth of a country. Though risky as an undertaking, it presents a challenging environment to test and implement technical solutions which can bring personal and professional satisfaction. Being an entrepreneur lets one build a career that is aligned with one's interests and helps in continuous personal growth in the field of work. An entrepreneur has the freedom to work with like-minded people toward a common goal.

6.9. Popular software packages in Civil Engineering

Programming languages and software packages are used to model, analyze, and design complex civil engineering processes and systems. Table 1 provides a list of a few important and commonly used software and programming languages related to different fields of civil engineering.

Table 1: List of software and programming languages relevant to the field of civil engineering.

Field	Name(s)	Purpose
Engineering drawing	AutoCAD.	Used for engineering drawings in 2D and 3D.
Structural Design	STAAD Pro, SAP-2000, ETABS, Revit.	Used for structural design and analysis.
Bridge design	MIDAS Civil.	Used in bridge analysis and design.
Hydraulic modeling	HEC-RAS, Delft3D, MIKE 21C.	Used for the simulation and study of hydraulic flow parameters (like velocity, depth, etc.) in an open channel.
Hydrological modeling	HEC-HMS, ArcSWAT.	Used for the simulation and analysis of hydrologic parameters (like soil moisture, infiltration, etc.)
Sewer and drainage modeling	SWMM, SewerGEMS.	Used to design and analyze sewer and drainage networks.
GIS/Geospatial tool	QGIS, ArcGIS, ERDAS.	Used for geospatial data visualization, interpretation, and analysis.
Traffic modeling	PTV Vissim.	Used for traffic modeling and simulation, traffic planning, designing transportation strategies, etc.
Transportation management and data analysis.	TransCAD.	A GIS tool specially designed for application in the field of transportation engineering. It is used to visualize, manage, and analyze transportation data.
Pavement analysis	KENPAVE, IIT Pave.	Used for the analysis of rigid and flexible pavements. It can calculate stress, strain, and deformation in pavements.
Construction Management	Microsoft Project.	A powerful tool used in construction project management for optimum resource allocation, scheduling, etc.
Programming Languages	Python, R, FORTRAN, MATLAB.	They are programming languages used in the field of engineering. Python and R are used for data analytics and modeling. FORTRAN is used for numerical modeling.

		MATLAB is a high-level programming language designed for engineering and scientific applications.

References

1. *Lexico*. (2022, 07 08). Retrieved from https://www.lexico.com/definition/plagiarism

2. Lindsell-Roberts, S. (2001). *Technical Writing for Dummies.* Indiana: Wiley Publishing, Inc.

3. Minto, B. (2009). *The Pyramid Principle: Logic in Writing and Thinking.* Pearson Education Limited.

Dynamic QR Code for Further Reading

1. Python numerical Methods https://pythonnumericalmethods.berkeley.edu/notebooks/Index.html	Python Book
2. Scipy Lectures https://scipy-lectures.org/	Scipy Lectures